THE WINNING STRATEGY

1. Do not harbor sinister designs.
2. Diligently pursue the Path of Two-Swords-as-One.
3. Cultivate a wide range of interests in the arts.
4. Be knowledgeable in a variety of occupations.
5. Be discreet regarding one's commercial dealings.
6. Nurture the ability to perceive the truth in all matters.
7. Perceive that which cannot be seen with the eye.
8. Do not be negligent, even in trifling matters.
9. Do not engage in useless activity.

THE BOOK OF FIVE RINGS
Herein Lies the Secret of Success in Life

Ed—
I would it.
Diane

The Book of Five Rings
(GORIN NO SHO)

by Miyamoto Musashi

Translation and Commentary by
Nihon Services Corporation:
Bradford J. Brown,
Yuko Kashiwagi,
William H. Barrett,
and
Eisuke Sasagawa

BANTAM BOOKS
TORONTO • NEW YORK • LONDON • SYDNEY • AUCKLAND

THE BOOK OF FIVE RINGS
A Bantam Book / May 1982
8 printings through July 1988

Poems on pages 29 and 101 by Lao Tzu from The Way of Life,
*as translated by Raymond B. Blakney; copyright © 1955 by
Raymond B. Blakney. Reprinted by arrangement with* The New
American Library, Inc., *New York, New York.*

ISBN 0-553-27096-6

Published simultaneously in the United States and Canada

---

*Bantam Books are published by Bantam Books, a division of
Bantam Doubleday Dell Publishing Group, Inc. Its trademark,
consisting of the words "Bantam Books" and the portrayal
of a rooster, is Registered in U.S. Patent and Trademark
Office and in other countries. Marca Registrada. Bantam
Books, 666 Fifth Avenue, New York, New York 10103.*

PRINTED IN THE UNITED STATES OF AMERICA

KR   17 16 15 14 13 12 11 10 9 8

To Peter Williams,
founder of Nihon Services Corporation.

Contents

HI NO MAKI
("The Fire Book")

KAZE NO MAKI
("The Wind Book")

KŪ NO MAKI
("The Book of Emptiness")

Translator's Note

In presenting this translation of, and commentary on, Miyamoto Musashi's *The Book of Five Rings,* it is our intention to provide you with a basic familiarity with the actual content of the original Japanese work, and by means of a commentary, a basic understanding of that work, and how it can apply to your own life.

Anyone who speaks or reads more than one language will agree that the best way to understand any material is to read it in its native language. There are shades of meaning that are always lost during translation. That is the nature of language. The second best way is to have a background, a context in which to consider the translation and further, to provide an explanation of unfamiliar concepts.

Generally, we have supplied background information and commentary as the subject matter comes up in the text, and such information appears prior to each passage or related groups of passages. There are some terms and concepts that require more lengthy investigation, and these are identified and explained in the Introduction just prior to the beginning of the text itself.

Since the intention of this presentation is to familiarize you with the material, we have refrained from adding copious footnotes or making the text less readable in any other manner. It is suggested that you read the comments first, and then read the passage. Also, where necessary, we have retained the Japanese word in the text, but this word is always discussed prior to its appearance in the text. This makes for smoother reading and permits some degree of individual interpretation. Also, there are some instances where material in brackets has been inserted in the text. This is not material found in the original text; it is there to enhance your understanding.

There are many sources available to you if you wish to delve deeper into any particular aspect of Japan or Japanese culture that you read about here. Our intention is not to make an exhaustive study of Japanese culture in the last four hundred years. Rather, we are investigating one aspect of Japanese culture, and focusing on what effects it can have on your life today.

—Bradford J. Brown for
Nihon Services Corporation, November,
1981

Introduction

The period in which Miyamoto Musashi grew up, the late 1500's and early 1600's, was one of great social change in Japan. For centuries, feudal lords had clashed over wealth, power, and control of parcels of land. Just before Musashi was born, the *Shōgun,* a military dictator, was able to inhibit the internal conflicts and unite the country. Subsequent *Shōguns* continued this trend, imposing rigid, bureaucratic rules over all of Japan. The strict class distinctions Musashi discusses were solidified during this period. Many *samurai,* as leaders and soldiers of the provincial armies and retainers of the feudal lords, became *rōnin,* or masterless *samurai,* as their armies were disbanded. They literally roamed the country; the two swords stuck in their belts as symbols of their status. Because there were effectively no more wars to fight, a *samurai* from one school of swordsmanship would often challenge a *samurai* from a different school to a match, to see whose technique was better, and of course, to test their blades.

Musashi, born a *samurai* in 1584, was such a *rōnin.* Beginning at the age of thirteen, Musashi had bouts with over sixty swordsmen before reaching his thirtieth birthday. It is said he never lost in these individual encounters. He fought with the long sword, the short sword, wooden swords, oars and fence posts; just about anything available.

Not much is known about Musashi between his thirtieth

and fiftieth years. It was during this time that he began to reflect on his life, and to formulate what he calls his *Heihō*. What is known is that Musashi became a skilled craftsman, sculpting in wood and metal, as well as a calligrapher. He was a prolific painter of *Sumi-e* paintings; *Sumi-e* being an expression of *Zen*, using a brush and ink to describe scenes of nature.

It has been said that Musashi met a *Zen* monk named Takuan, whose influence on Musashi was profound. Takuan, himself a swordsman, wrote a letter to a student of his, one Yagyū, (another famous swordsman), concerning the relationship of *Zen* and the art of swordsmanship. This letter is referred to as the *Fudōchi Shimyō Roku*; it describes in greater philosophical detail that which Musashi writes pragmatically about in *The Book of Five Rings*.

This twenty-year period, then, for Musashi was a time for reflection, perfection of technique, and the seeking of other disciplines to broaden his own knowledge.

At about the age of fifty, Musashi "came face to face" with the realization of his *Heihō*. He had become aware of the forces that shaped his life. Several years later he wrote out the technical principles of his *Heihō* as they related to swordsmanship. This document is called the *Heihō Sanjūgo Kajō, The Thirty-five Articles of Heihō*. Two years after that, in the form of a letter to his student Terao Magonojō, Musashi wrote out *The Book of Five Rings*, as a textbook guide for the attaining of his *Heihō*. He died shortly after completing the book, in 1645.

The Book of Five Rings is not merely a theoretical dissertation. Musashi derived his method and its philosophical and psychological underpinnings from practical experience. Musashi's intention was that this approach to life be applied to everything you do everyday, all the time. It is sufficiently detailed, so that with a little imagination, the method that was realized 300 years ago is just as valuable today.

The book itself is divided into five chapters. As you will discover, the primary philosophical point of *Heihō* is that all things are dependent upon other things. These five areas of discussion are to be considered as each existing in relation to all the others.

The first and fifth chapters are the beginning and the end. The Earth Book sets the basic groundrules for understanding, providing the context, the framework in which to view and understand the rest. The last chapter, the Book of Emptiness, is the culmination of all that comes before; it is the final key to understanding.

Books two, three, and four—Water, Fire and Wind, respectively—are practical in nature, giving you the technique itself. The Water Book, (which tells you about Musashi's technique), when juxtaposed with the Wind Book, (which describes the techniques of all the other schools), provides meaning to the saying, "To know the enemy, know yourself." The Fire Book gives the strategy with which to combine the technical knowledge gained from the Water and Wind Books.

As it was in Musashi's time, *Zen* is an inherent part of the Japanese culture today. Its influence can be seen in art, in craftsmanship, in war, in sports, in business, and in the daily activities of the Japanese people. For this reason, it is important to gain at least an idea of what *Zen* is supposed to be, what it represents, and how it affects one's life. We have provided a mere glimpse into what *Zen* may be in the following section.

Further, to provide a more familiar environment in which to consider *The Book of Five Rings*, both the subjects of *Bushidō* and of *Heihō* have been introduced in their own sections. They provide some background for a more general understanding of both Musashi and his Way.

If you ask almost anyone if they have heard of *Zen*, the answer is usually yes, but when you ask them what *Zen* is, the answer usually varies greatly. The word *Zen* appears in book titles, and in university course outlines. Books are written about experiences where *Zen* has allegedly played a part, and groups are started based on what *Zen* is believed to be. The idea of *Zen* is popular. But the practice of *Zen* is not so popular, in the sense that relatively few people actually practice *Zen*—consciously, that is.

Saying that one does *Zen* is ultimately the same as saying one breathes or works or eats. However, the fruit of *Zen* practice is an awareness, a heightened sense of feeling and sensitivity that one develops in doing what comes naturally.

As a frame of reference, consider *Zen* both as a discipline, a very strict and demanding discipline, and as an attitude, or outlook or approach to life. *Zen* has no philosophy of its own: it is simply a method of learning, of observing, and of experiencing what you do every day, anyway. *Zen* is not a religion, in the sense of a "God-fearing" religion. It is a set of values or beliefs, and a way of life for some. A practitioner of *Zen* does not give up his particular faith. *Zen*, if anything, makes one's convictions stronger, whatever they are.

Zen is a philosophy of will power—one forges an iron will and an indomitable spirit. There are no special doctrines or teachings. There is no deity to depend on or to assuage one's conscience. You are saddled with the responsibility for yourself.

Zen is based on a spirit of self-reliance. You must learn to depend on yourself. But this is no simple statement. Think of the criteria you consider before you depend on somebody else. Do you meet the criteria? You now know that profound honesty is required in *Zen* practice.

There is no result in *Zen* practice. That is not the point. It is the effort you make to improve yourself that is measured.

There is a strong work ethic associated with *Zen*. The teacher, or master, does not sit idly by, watching the work being done. The teacher and the student share the burdens. Each does his share of the chores; each receives the same benefits from his work. There is the story of the aged *Zen* master whose students felt that he should not be tilling the soil as they did, for he looked weak. They hid his tools so that he could not work. When they sat down for the evening meal, the master would not eat. And so it went for three days. Finally the students retrieved the tools, and upon thanking them for finding his tools, he said, "No work, no food."

As one way of practicing *Zen*, one sits in a cross-legged position called *Zazen*. With your hands relaxed, your arms relaxed, your shoulders relaxed, you begin to breathe in a controlled, rhythmic breathing that quiets your activity. The breathing is deep, nourishing you with oxygen; your nostrils flare and your abdomen, the *tanden*, perhaps for the first time in your life, breathes for you. Thoughts fly through your head. You are admonished to concentrate on your breathing, to measure it, feel it, be aware of it. The thoughts increase: the day's activities, your problems, your fears, a million things fly by. You snap back into where you are, and what you are doing. The more you try to concentrate on one thing, the more it slips away. Later, when you become experienced, you will not try to stop your thinking. You will let it stop by itself. You will let it go. You will realize that nothing outside of you causes you trouble or anguish or fear or guilt or doubt. When a thought occurs to you, and you spend time thinking about it, it is said that your mind has stopped at that point. This "stopping mind" is the heart of the problem. When your mind "stops" to question or decide or judge—when you are concentrating on that, you lose track of what is still going on.

But the flow does not stop when you do. It just passes you right by. When the mind is stopped for too long, it causes you not only to falter, but maybe to lose days in responding to a problem. In *Zen,* an instant is too long for your "stopping mind." When you have guilt or fear or anxiety or regret or nervous anticipation, you are living in the past or the future. You must live now, in the present. In this moment! But how do you come to realize this?

In a famous *Zen* story, Nan-in, a *Zen* master, was visited by a university professor who wished to know all about *Zen.*

Nan-in served the professor some tea. He poured the cup full, and then continued to pour. The professor watched the overflow until he could no longer restrain himself. "Stop it! No more will go in!"

"Like this cup," said the *Zen* master, "you are full of your own ideas and speculations. How can I show you *Zen* unless you first empty your cup?"

The initial concept with which *Zen* training breaks open your doors to understanding is "Beginner's Mind." The *Shōrinji* (a school of *Zen*) tells you to learn as "plain and naive as an infant."

The mind of an infant is empty; it is fresh. It has no preconceived ideas; it sees things as they are. It is free from the habits of experience and therefore open to all possibilities. The infant has no thoughts of achievement, and makes no demands. It makes no judgments, no distinctions. The infant lives in the absolute present. Above all, because he does not put one before another, the infant is compassionate.

Zen is a practical discipline. As opposed to the Buddhist schools, in *Zen*, enlightenment can be achieved in this lifetime. *Zen* wants you to act now, to experience this moment right now, directly. The effect of such action is to give you the power to cope. That is important in today's world. Furthermore, the resiliency that is developed in one's practice allows for appropriate responses. Coping appropriately is a key concept.

Zen teachings are concerned with the practitioner attaining an intuitive experience. The nature and context of the experience are irrelevant. As long as there is no intrusion by the intellect, as long as the experience is immediate (or im-mediate as the famous *Zen* scholar Dr. D. T. Suzuki writes); that is, when there is no mediation by the intellect, this is an intuitive experience. It is understanding without words. It is to apprehend the situation clearly and to see it for what it is, and not what you think it is. The intuitive experience is quick and without hesitation. There is no resistance. This intuitive moment is not instinct. It is spontaneous, but not instinctive. The body is allowed its own wisdom, and is completely free from any mental steering. It is a natural action, like a bamboo stalk that bends under the weight of snow until the snow is thrown off, leaving the stalk where it was before. The snow is gone, but there was no conscious or thoughtful purpose involved in the activity.

So, reality must be experienced as it is. Verbal instruction is futile. Does any word or even any thought accurately describe the feeling of being in love, or the heart-stopping fear of imminent death? A *Zen* exercise to make one realize the futility of language is called *kōan*. The *Zen* master Hakuin asks the question, "What is the sound of one hand clapping?" the *kōan*, or "riddle," is offered to point out that at times there are no intellectual answers. Intellectual understanding is not always sufficient to effectively communicate. In *Zen* practice the *kōan* is used to measure the level of concentration or intuitive perception one has attained.

It is understood by the *Zen* mind that the senses cannot grasp reality from one viewpoint. For example, the *Zen* garden at Ryōanji, a *Zen* temple near Kyoto, appears as a few rocks and some sand. The garden begins to make some sense when you realize that from your vantage point you cannot quite see all the rocks. You might also notice that you are picking out only the rocks to look at. Is not all that sand just as important? What if it were all rocks? Would you be trying as hard to see all the rocks?

When looking at a "*Zen* picture" (*Sumi-e*), one sees the brushwork amid an otherwise empty background. Most other paintings cover every inch of the canvas. The *Zen* painter wants you, in one instant, to feel and understand, in the deepest sense of the word, what you are experiencing. You have to be aware. When the light goes on, one says, "The light is on." But are you aware of the absence of darkness? Have you considered all sides of the situation, in your observation of it? *Zen* prepares you to look at the spaces around the rocks.

Dōgō, a *Zen* master, observed, "If you want to see, see right at once. When you begin to think, you miss the point." (Read this again for there are two points you may want to understand.) The idea is, of course, not to intellectualize the experience. Satisfying sex or a good joke are just two examples of the point to be made. To know and to act are one and the same. Be natural. Just do it.

Think of any activity that you perform that you are good at. It can be investment banking, cooking, tennis, speechmaking, organizing, or anything! What is it that makes you good at it? Is it your training, or the tools you use? Or is it the experience you have accumulated in doing it? It is all of these things, in varying degrees, but the missing element is the crucial one. Your attitude, your approach, the sense of confidence and purpose (no hesitation) you bring to your activity is what people observe when they say you are "good at it." *Zen* is a practice for life; in *Zen* first comes the technique, practiced so many times that it is forgotten. Then you begin to use it. It is when you do not think about it anymore that you do it so well. *Zen* is no more than that. But it is reaching that state that the training is all about. The professional dancer who makes it look easy has trained constantly and endured great pain. The tennis pro who flies around the court, making impossible shots, does so not because of any superhuman qualities but because he has practiced and practiced, as the dancer has, until the movements are internalized. There is no longer any conscious direction in the movement. When you marvel at the way someone whips up a dinner for ten on short notice, or the way someone makes an impromptu speech, you are marveling at the same thing—the approach, the confidence, the naturalness of the behavior. There was no time to prepare, no time to think, no time to hesitate. There you are. *Zen.*

Zen is a social philosophy as well. A goal of *Zen* is to realize your potential as a human being. The "self" is understood not only as an individual, but as a member of the community of individuals (society). *Zen* stresses self-perfection, and in so practicing, one tends to be more aware of one's place in the world, not in the sense of "better," but more in the sense of "sufficient." The teaching is to be applied to daily life. That is the point. One's personality becomes one of purpose, in the sense of dedication or perseverance. If you have ever tried really hard to do something, and accomplished it, you know the feeling. It is a sense of self-satisfaction. You carried out your plans. You accomplished your goal. You felt good! But if you can do that yourself, why *Zen*?

Zen disciplines you to feel that way every day, in everything you do. You become more observant, more aware, more sure, and more confident. The effect of this change is to reduce your ego: the more confident you are, the less likely you are to boast or brag. But if your confidence is in only one activity that you perform, your ego (your personal advertising agency) will boost your confidence for you in other ways. All the cheerleading in the world will not help you win the game. If it is *you* who you have confidence in and not just your skills, if you are satisfied with yourself, you will be humble and quiet and peaceful. You will not have the *desire* to show off. When you are confident, truly at peace, you are also benevolent. Your ego does not keep you from reaching out to lend a helping hand, or from being compassionate. Your ego is not there to interfere. You do it because you feel like doing it, and there is nothing to stop the feeling.

There is the active way and the passive way to practice *Zen*. This refers only to your actual practice sessions, for one should always, at every moment, be practicing. The "passive way" has been described as *Zazen;* seated, contemplative concentration, and relaxation.

The active way is what *The Book of Five Rings* is about. The actual form of expression is left up to the individual. The task of *Zen*, as the late Suzuki-Rōshi, one of the first proponents of *Zen* in the United States, said, "is to make you wonder, and to answer that wondering with the deepest expression of your own nature . . ."

Sumi-e ("brush painting"), *Ikebana* ("flower arranging"), and *Haiku* ("poetry") are classical Japanese expressions of *Zen*. More well-known outside of Japan are what is commonly termed the martial arts. The martial arts that are based on *Zen* practice are *Kendō*, *Iaidō*, *Jūdō*, and *Shōrinji Kempō*. At first, it may seem incongruous to connect what has heretofore been described as a relaxing, peaceful, contemplative, and benevolent philosophy with the martial arts. But think, when is there ever a time that an immediate and appropriate response is more necessary than in the context of battle?

In the active expressions of *Zen, Zazen* is practiced for the training of the mind. The physical activity you do is the training for the body. There is a saying, *Ken-Zen Ichi Nyō,* ("Body and Mind, Together"). As was said earlier, first the technique is practiced so often that it is internalized and "forgotten," and then one learns to use it. Mastering the technique is mind over body; discipline, hard work, forcing the body to accept the rules and the pain and the utter exhaustion of constant practice, until the body *learns.* Using the technique is body over mind; the body just does what it knows now how to do.

You see, when you walk or eat or sleep—your body knows how to do that: your mind is not involved. To understand the point here, the next time you approach a flight of stairs, using the same speed you always use, walk up or down but this time look at each step as you proceed. When you are tired you sleep, and when you are hungry, you eat. You usually have no trouble eating. But have you ever had the experience of having a meal with someone you are really trying to impress? It is simply amazing how the food seems to fly into your lap, and it is magical how the water glass keeps falling over.

When you eat, just eat. When you walk, just walk. When you sit, just sit. Your body knows how to do that already. When you play tennis, just play tennis. Approach the situation for what it is, and nothing more: whether you like it or not is irrelevant.

A *Zen* story illustrates this: Two monks were traveling in the rain, the mud sloshing under their feet. As they passed a river crossing, they saw a beautiful woman, finely dressed, unable to cross because of the mud. Without a word, the older monk simply picked up the woman and carried her to the other side.

The younger monk, seemingly agitated for the rest of their journey, could not contain himself once they reached their destination. He exploded at the older monk. "How could you, a monk, even consider holding a woman in your arms, much less a young and beautiful one. It is against our teachings. It is dangerous."

"I put her down at the roadside," said the older monk. "Are you still carrying her?"

This story, a favorite one, brings into focus a resolution of the seeming contradiction between a master of *Zen* and a master of war being one and the same person. Our friend the old monk picked the woman up and put her down. That's all. No mediation by the intellect. He just did it. His mind, other than to work his muscles, was not a part of the experience. Yet the rules of his order prohibited his behavior. Did he do the right thing? Was it appropriate?

Morality is judged by intention, that is, subjective intention. It is clear that to take a life is immoral. But in the context of a volcano erupting and killing hundreds of people and destroying livestock and so on, it is hardly appropriate to judge this as immoral. Why? Because we do not think of the volcano as intending to do the damage. On a grander scale, it is not appropriate to make any moral judgments of any act of Nature for the same reason. To question whether any act of Nature is appropriate or "right" is equally inappropriate. Nature (its manifestations) just happens. It just does.

The old monk just did it.

And so we come to a resolution of the problem. Our master of *Zen* acts with his *Zen* mind when he lifts his sword. When he strikes, he is doing only what his body knows how to do. He sees it, does it, and drops it. No morality involved. No intention. It is not the act of cutting that is immoral. A kitchen knife can cut apples or can be used for more sinister designs. It depends on the intention of the user. *Munen musō* is a *Zen* phrase meaning, roughly, "where there is no intention, there is no thought of doing." Something else moves the sword.

This is the source of the often misunderstood concept of the martial arts. The martial arts have been described as "self-defense," and are thought of in terms of peace, benevolence, humanity, restraint, and so on. But the question arises as to how one can style these activities as "self-defense" when, in fact, the techniques are mostly offensive, first of all, and secondly, they are violent and aggressive actions.

There is a difference between a boxer who wins, and a boxer who continues to beat his adversary to a bloody pulp long after the bell has sounded. There is a difference between one who provokes a fight, and the one who is provoked. Both are fighters, but that is not the issue. Both the one who is provoked, and the boxer who wins fairly and with respect are not blamed for their behavior. But there is a point after which society condemns even the protector for his actions.

That is the line that the *Zen*-trained martial artist does not cross. He truly acts only in response to aggression. He does not seek it out. When made, his responses are non-resistant and non-violent. He is a man of peace, content to paint his paintings, arrange his flowers, and practice his swordsmanship, in all ways constantly refining his technique.

When called upon to act, he simply acts. When he paints, he paints. When he uses his sword, he simply uses his sword. When he is pushed, he does not push back. He lets whatever it is go right past him. His response is purely defensive. It is also decisive.

We have tried to introduce *Zen* to you by touching on those considerations that you might find useful in your daily life, and those that will help you to put Musashi in context with what you will read in *The Book of Five Rings*.

Zen is not complicated; on the contrary, it is rather simple. But it is a demanding discipline with a rigorous physical practice method. It makes you strong physically, firm in your actions, and resolute in your attitude. Musashi practiced *Zen* without a teacher. Life was his teacher. He learned his lessons well. You must understand Musashi in this context.

BUSHIDŌ

The word *Bushidō* is written with three characters, *bu* ("military"), *shi* ("man"), and *dō* ("way"). Hence, *Bushidō,* the way of the warrior; otherwise known as the *samurai* code of chivalry. This term describes the principles of honor and loyalty followed by the *bushi,* members of the military class that ruled feudal Japan. Miyamoto Musashi was a member of this military class.

The *bushi,* the highest status group in feudal Japan, accounted for only 5 or 6 percent of the population. This social group included the *Shōgun,* about two hundred "feudal lords" *(daimyō)* and their retainers, from top level administrators in the *Shōgun's* government, down to foot soldiers and gate guards. They were sharply differentiated from the other main social classes: the peasantry, who constituted about 80 percent of the population, and the artisan and merchant classes. The *bushi* lived on rice stipends drawn from the taxes paid into the lord's treasury by the other three social classes.

The code of the military class, *Bushidō,* was a synthesis of borrowings from three main sources. It borrowed stoic endurance, and scorn for suffering and death from *Zen,* worship of country from Shintō, and the social ethic of the five relationships from Confucianism, the most important being that between lord and retainer (the retainer owed his lord uncompromising loyalty).

The ideal *samurai,* in preachment if not always in fact,

was loyal until death to one master in whose service he was always willing to sacrifice his life without a moment's hesitation, "like the cherry tree sheds its blossoms."

The *samurai*, as a professional warrior, was proficient in the use of many weapons: the bow, the Japanese "spear" *(yari)*, the Japanese "halberd" *(naginata)*, and firearms, but the sword was his weapon of choice. The *samurai's* swords were the badge of his class. It took years of arduous physical and mental training to become proficient in their use. *Bushidō* emphasized constant physical training to maintain and improve techniques of swordsmanship and austere, *Zen*like discipline to develop the character, the confidence, and inner self-control needed by the *samurai* to face an opponent's blade in battle to the death, without flinching.

Bushidō is a philosophy that teaches patience. In single combat the *samurai* swordsmen stood face-to-face within striking distance of each other and waited for the opponent to make the first move. The weaker man, no longer able to bear the strain of waiting for a blow from the sharpest blade known to man, would eventually strike the first blow. But, the instant he made his move, the other man also would move, not to defend himself but to attack. This kind of confrontation, which rewarded a moment's relaxation with instant death required awesome patience and concentration, a kind of discipline that can only be acquired after years of training under the guidance of a master. In time, this code of ethics with its stress on patience, frugality, and constant self-improvement, permeated all levels of Japanese society. It became a part of the social ethos of Japan.

The kind of patience demonstrated by the *samurai* trained in *Bushidō* can still be found in modern Japan. For example, when an American businessman arrives at the headquarters of a major Japanese enterprise in Tokyo to negotiate a contract, he will be ushered into a conference room where he will be politely seated and offered a cup of Japanese green tea by the Japanese negotiators sitting across the table from him. After the tea is served, the Japanese will simply sit, wait, and say nothing. They are waiting for the opponent to make the first move. The American, inexperienced in the techniques of Japanese negotiators and feeling very uncomfortable, will attempt to open negotiation (or get down to business, as he might put it). With this technique, the Japanese gain the upper hand. They have learned the opponent's position without revealing their own. The experienced foreign negotiator who knows his Miyamoto Musashi *Bushidō* or *Heihō* will just silently sip his tea and wait, no matter how long, for the Japanese hosts to open the talk.

An acquaintance with the teachings of Miyamoto Musashi then can provide insights into the Japanese way of doing things that can be very useful to the modern businessman no matter what nationality he deals with.

HEIHŌ

In Musashi's view, *Heihō* literally means the path to enlightenment. It is not enlightenment itself. While enlightenment or fulfillment or success in accomplishing your objectives may be the goal, *Heihō* is merely the way to get there—Musashi's way to get there. While Musashi is convinced that using his *Heihō* will insure success, it is not the success, or the goal, that is the point.

It is the way you do something, how you do something, rather than what it is you actually do, that is addressed by *Heihō*. *Heihō* is what it takes to develop the right outlook, the right attitude, that frees you to be successful. Remember that success here is not the goal; the goal is the correct application of Musashi's *Heihō* to your activities.

Musashi was a masterful swordsman. In the context of early seventeenth-century Japan, this meant he was a soldier, a warrior. He of course, was a *bushi* and as a warrior, he faced life-and-death situations constantly. It is said he never lost an encounter. From knowing his own superiority, he believed that there was, first, a level of understanding of life that could be reached whereby one became invincible, and second, that the way to this understanding was through what *he* had gone through to be invincible. He called the way to this understanding *Heihō*.

The Book of Five Rings is about Musashi's *Heihō*. There are other paths to success, obviously. But the context of Musashi's way of achieving success is one of a "no-holds-barred" situation—literally one of life and death. You cannot make a mistake. You are faced with a person intent on slicing you into ribbons with a four-foot razor blade. The usual psychological ramifications of such a stressful situation are fear, anxiety, panic, confusion as to what to do, and so on. The physical manifestations are sweating, nausea, trembling, and weakness, not to mention paralysis. Such reactions are hardly appropriate. You will surely be chopped in two.

Heihō prepares you for such an encounter. It gives you the tools to use to take the initiative, and to win. For those who are disposed or required to live in a world where winning and losing are the primary considerations, Musashi's *Heihō* will show you the way to win.

For those whose only interest is to be fulfilled, Musashi's *Heihō* will set you free.

PROLOGUE

Commentary to the "Prologue"

In a prologue to the work itself, Musashi explains what transpired in his life prior to his attaining what he describes as a deep understanding; what he calls *Heihō*.

He makes reference to the fact that having discovered *Heihō*, he has no need of any teacher. It was usual to have a teacher or master when studying any art or skill. But Musashi's *Heihō* is the way to do *anything*; it is an attitude about life itself, therefore he needed no teacher. For example, Musashi became a very skilled and prolific *Sumi-e* painter, without a teacher to guide him.

Sumi-e is a *Zen* painting technique. Rather than using a model, the painter draws freely from his memory and imagination, using a brush and ink. It is a direct action with no hesitation. Bold, confident strokes are used. Dr. D. T. Suzuki's analogy is the best: "*Sumi-e* reflects life . . . when painting, the least stroke made over another becomes the error visible after the ink has dried. In life, one cannot take back what has been done. *Zen* teaches that life must be seized at the moment, not before or after."

It was also the accepted practice to quote from previous works or from Scriptures to support one's views. Musashi does not do this. He also advocated a two-sword technique; this was also against the prevailing methods. Further, he advocated a "religion of the self," rather than the accepted Buddhism or other religions practiced at that time.

3

Zen is a philosophy of self-reliance. *Zen* practitioners are not often followers. They are leaders. In the words of the writer and editor of *Zen* works, Paul Reps, instead of being followers, *Zen* practitioners aspire to place themselves in the same responsive relationship with the universe as did Buddha and Jesus, so that they may experience it firsthand. The Buddha said, "Look within, thou art the Buddha." Jesus said, "The kingdom of Heaven is within you." This was Musashi's insight, as well.

Prologue

I wish to put down in writing for the first time that which I have been disciplining myself in for a number of years, and to which path of *Heihō* I have given the name *Niten Ichiryū*. [Two Heavens—as—One School. Sometimes referred to as *Nitō Ichiryū*, or the Two Swords—as—One School.] It is the early part of the tenth month of the twentieth year of the *Kanei* era [1643]. I climbed Mount Iwato in the domain of Higo in Kyūshū to worship Heaven, [reference is to Shintō, the native Japanese religion of ancestor and nature worship] pray to Kannon, [Buddhist Goddess of Mercy], and to honor the dead. [Pay homage to departed teachers and elders.] I am Shinmen Musashi no Kami Fujiwara no Genshi, born a *bushi* of the domain of Harima [now the southwestern part of Hyōgo Prefecture, near Kobe on the main island, or Honshū]. I have reached the age of sixty.

I have devoted myself to the path of *Heihō* since my youth, and had my first match at the age of thirteen. This bout was against a swordsman named Arima Kihei of the *Shintōryū* school, and I was victorious. At the age of sixteen, I won against a formidable swordsman named Akiyama, of the domain of Tajima [now the northern part of Hyōgo Prefecture]. At the age of twenty-one, I went up to the capital [Kyoto] and met with swordsmen from throughout the country and had a number of confrontations with them, and not once did I fail to be victorious. After that, I went to different provinces, and met with swordsmen of the various schools, and had over sixty bouts with them. I did not lose even once. This was between the ages of thirteen and twenty-eight or twenty-nine.

After passing the age of thirty, I reflected on the road I had been traveling, and came to the realization that I had won not because I had attained the full secrets of swordsmanship, but perhaps because I had natural ability for this path, or that it was the order of Heaven, or because the other schools of swordsmanship were deficient. After that I tried to attain a deeper understanding, and as a result of disciplining myself day in and day out, at about the age of fifty, I came face-to-face with the true path of *Heihō*.

Since then, I have passed the time without needing any particular path to follow. [I need not look for another.] Having become enlightened to the principles of *Heihō*, I apply it to various arts and skills, and have no need of any teacher or master.

Similarly, in writing this book, I have not quoted from the ancient words of Buddhism or Confucianism, nor have I used the ancient military chronicles about military tactics. In the light of the path of the Heavens and Kannon, at the hour of the tiger [four A.M.] on the night of the tenth day of the tenth month, I have taken up my brush and begun to write.

CHI NO MAKI

("The Earth Book")

Commentary to the "Earth Book"
CHI NO MAKI
("The Earth Book")

In the *Chi no Maki, ("The Earth Book")*, Musashi discusses the state of things as he sees them, especially the commercial activities of those who would sell their sword technique to make a living.

He uses the analogy of a master carpenter to discuss the important points of *Heihō* in their purely military sense, that of commanders and soldiers.

An interesting homily appears in this section: "The teacher is the needle and the disciple is the thread." In typical *Zen* training, one is guided by a master. The master, or teacher, guides the student to the student's personal understanding of whatever is being done, with enlightenment as the ultimate goal. The student is part of the resulting awareness. The master is not. But the master is *necessary* to make the student understand. Someone has to tweak the student's nose!

A summary of the chapters (each book) appears next, but within each summary is valuable information. You must remember to carefully read each word and ponder the meaning. Let your mind be open to all possibilities of meaning.

Musashi discourses next about weapons and their use. In his terminology, the terms for swords are in relation to their size. A *tachi* is "a long sword," and a *wakizashi* is

9

"a short sword." A *katana* is also "a long sword," but is a bit shorter than "a long sword," *(tachi)*. Therefore, we have referred to a long sword as a *tachi*, to a sword as a *katana*, and to a short sword as a *wakizashi*.

● *From one thing know ten thousand things:* Knowing the principles of *Heihō* will give you the understanding of anything. The real message here is that being adept (in the sense of *kū*) at anything gives one the opportunity to take what one knows and apply it to other things. Musashi constantly exhorts you to not just learn, but to apply your learning to *whatever* you do.

● You will see the word *heihōsha* in the text. Here *heihōsha* is merely "a swordsman," as opposed to one who is a practitioner of *Heihō*.

● When Musashi discusses the virtue of "the long sword" *(tachi)*, he means the virtues of the *bushi*, or *samurai;* the code of honor and loyalty, and benevolence and humility, as represented by the sword.

● The *Noh* dancers refer to *Noh* plays, or the "dance-drama theater of Japan."

● Musashi talks of habit and complacency, and indicates that the Middle Way, or the Way of no extremes, is the better way. The "Middle Way" is the name given to the *Madhyamika* school of Buddhist thought which evolved the final version of their conception of reality. This is addressed in the last chapter, "On Emptiness."

● Read the section on rhythm with care. Recall it when you read the last chapter ("On Emptiness"). What is sought is harmony among all things. This harmony is expressed by remembering context and responding appropriately within that context.

● At the end of the chapter, Musashi says that certain results will obtain from the practice of *Heihō*. These are the very same that the *samurai* are dedicated to.

Introduction to the "Earth Book"
CHI NO MAKI
("The Earth Book")

First of all, *Heihō* is the way of the warrior class. Specifically, a commander is the one who must execute this way; foot soldiers must also have an understanding of this way. However, in today's world, there are no warriors who have firmly attained an understanding of the path of *Heihō*.

In speaking about paths, there is the path of Buddha, by which people are saved. There is the path of Confucianism [the learning of the scholars], the path of healing various illnesses for physicians, the path to teach *Waka* ["classical Japanese poetry"] for the poets, the path of the men of refinement [tea ceremony, flower arranging, music, and so forth], the path of the archers, and other arts and skills. People practice in their own path, to their personal satisfaction. But there are few who are inclined to devote themselves to the path of *Heihō*.

It is the warrior's way to follow the paths of both the sword and the brush [pen]. Even if he has no natural ability in these paths, a warrior is expected to do his share to the best of his ability.

It is generally accepted that the way of the warrior is the resolute acceptance of death. [The question is only when and how.] But the way of death is not limited to warriors. Priests, women, and peasants may choose to die if obligation or avoidance of shame forces them. [In the decision to die in these contexts, there is no difference between warriors and others.] In order for a warrior to follow the path of *Heihō*, it is necessary to keep in mind that the essence of *Heihō* is to build an indomitable spirit and an iron will; to believe that you cannot fail in doing anything. The warrior must win in one-to-one combat or in battles involving large numbers, so that he can gain honor and fame for his lord [*daimyō*] and himself. This is accomplished by the virtue of *Heihō*.

There are some who believe that even if you master the way of *Heihō*, it will be of no use. The true path of *Heihō* is such that it applies at any time and in any situation.

HEIHŌ NO MICHI TO IU KOTO
("The Path of Heihō")

In both China and Japan, men who practice the path of *Heihō* have become known as masters of military tactics. Warriors must study these principles.

In recent times there are men who make a living styling themselves as swordsmen, but they teach only the standard techniques of swordplay. In recent times, the *Kashima* and *Katori* priests in the Hitachi domain have established their respective schools of sword techniques as the teachings of the gods, and tour the country teaching people. These are events of recent years.

Since ancient times, *swordsmanship* has been included among the ten skills and the seven arts as "profitable measures." Truly, it is one of the arts, but "profitable measures" are not limited to the standard techniques of swordsmanship alone. It is difficult to know the art of swordsmanship solely by means of the techniques of the sword. Needless to say, such swordsmanship cannot be expected to rival the principles of *Heihō*.

When we look at the world, we see various arts offered

12

as items for sale. Men think of themselves as commodities for sale. There is a trend for men to invent various tools and to sell those rather than their faculties. This thinking is like separating the seed from a flower and valuing the seed less than the flower.

Actually, this way of thinking causes them to color their technique and "show it off." Advocating this school's or that school's sword technique training hall, they try to gain profit by teaching or learning techniques. The result, to use a common phrase, is that "a little knowledge can be dangerous."

In general, there are four ways a man can make a living: warrior, peasant, artisan, and merchant.

● First, the way of the peasant, or farmer. Peasants possess various tools and spend their lives attending to the vagaries of the four seasons. This is the way of the peasant.

● Second is the way of the merchant. *Sake*-sellers buy the various ingredients, put them to use, and make a living by taking a reasonable profit.

Both have an occupation suitable to their station and make a living by means of taking profit. This is the way of commerce.

● Third is the way of the warrior. The warrior, in accordance with his aims, maintains various weapons and knows their characteristics and uses them well. This is the way of the warrior. Is it not an indication of superficial accomplishment of a warrior if he does not master various weapons and understand individual weapons?

● Fourth is the way of the artisan. To make his living, a carpenter must skillfully maintain various tools, and employ each of these tools skillfully. He must correctly draft plans with a measure and do the job properly. This is how he makes a living.

Let me explain military tactics by way of comparison with the craft of the artisan. I shall compare it with the carpenter because of the relationship with houses. We speak of the noble houses, the military houses, and the Four Houses. Houses perish and houses continue, but because we speak of schools, styles, and professions as houses, I shall employ the craft of the carpenter to explain military tactics. Because the word "carpenter" is written with the Chinese characters that mean "great scheme," and the principles of military tactics are also a "great scheme," I make a comparison with the carpenter. If you are thinking of studying the principles of military tactics, read this book and think deeply.

The teacher is the needle and the disciple is the thread. One must practice ceaselessly.

HEIHŌ NO MICHI DAIKU NI TATOETARU KOTO ("Comparing the Path of Heihō to that of a Carpenter")

If *Heihō* is to be compared to carpentry, the general is the master carpenter who knows everything about the nature of the carpenter's square and who is knowledgeable about the ways of the region and the particular ways of that household for which the carpenter is building the house. This is the path of the master carpenter. The master carpenter learns the measurements of the temples and the plans of the palaces, and constructs the houses which people use. The same is true for the master carpenter as it is for a commander of a warrior house.

In building a house, the lumber is sorted. Straight, gnarless lumber which is beautiful to behold is used for the external pillars; lumber with a few gnarls but which is straight and strong is used for inconspicuous pillars; lumber which may be a bit weak but which is gnarl-free and beautiful to behold is used for the threshold, lintel, doorways, and *shōji* ["sliding panels"]; lumber which is gnarled and slightly distorted but which is strong, is used in places appropriate for such lumber after careful study has been made for such places, thus resulting in a house which

14

is solid and long lasting. There is also lumber which is heavily gnarled, distorted, and weak, and such lumber can be used for scaffolding and later as firewood.

When the master carpenter is using other carpenters, he must know the degrees of ability of each and assign them to their tasks accordingly, some to the alcove [*tokonoma*; place reserved for a special scroll or art piece; the guest of honor sits before it], some to the doorways and *shōji*, some to the threshold, lintel and ceilings, and those who are not skilled to joist making, and the least skillful, to wedge shaving, thus making good use of the personnel in a way so as to increase productivity.

To accomplish a task quickly and to perform it well is not to be haphazard about anything; to know where and when to use who and what; to know whether or not there is incentive; to give encouragement and to know limitations; these are what a master carpenter keeps in mind.

The principles of *Heihō* are the same.

HEIHŌ NO MICHI
("The Way of Heihō")

The carpenter is like a rank-and-file warrior. For example, the carpenter sharpens his tools himself, maintains various tools, and carries them around in a box. He follows the instructions of his chief, whittles pillars and beams with an adz, planes the flooring and shelves, and also does open-work woodwork and chiseling. He measures correctly and carefully completes every corner [nook and cranny] with skill. This is the art of the carpenter.

Personally learn well the techniques of the carpenter. When you understand the plans well, in due course you can become a chief.

A necessary accomplishment of the carpenter is to have sharp tools; he must sharpen them in spare moments. The main task of the carpenter is to skillfully produce *zushi* [small shrines], bookshelves, tables, paper-covered lamp stands, chopping boards, and even pot lids with these tools. Warriors also must act in this manner. One must study this carefully. The necessary accomplishments of the carpenter are that his work not be warped and that joints are aligned. The work must be smoothed out with a plane and not polished haphazardly so as to disguise defects. It is important that the work does not warp afterward.

If you are considering learning the principles of *Heihō*, it is necessary to carefully consider the things written here one by one, and to study hard.

KONO HEIHŌ NO SHO GOKAN NI SHITATSURU KOTO
("On the Five Chapters of this Book on Heihō")

That which is accomplished by the five chapters of this text on *Heihō* is done by means of dividing *Heihō* into five aspects, each of which is taken up in a separate chapter, and in order to indicate the contents of each, these five chapters have been designated as *chi* ["earth/land"], *mizu* ["water"], *hi* ["fire"], *kaze* ["wind"], and *kū* ["emptiness"].

First of all, in the *chi* chapter, an outline of the paths of *Heihō* and an explanation of the way of thinking, or the philosophy of my school are given. If one does only sword technique, it is not possible to know the true path of *Heihō*. It is that which must be known from the large matters to the small matters, from the shallow matters to the deep matters. In order to lay down the groundwork for a straight path, this first chapter has been named *chi*.

In the second *mizu* chapter, using water as a model, the spirit is made to be as water is. Water alters its shape according to the shape of its container, be it square or round, and it can be a drop or an ocean. There is a green-blue color to water. Borrowing from its clarity, I have written about my style of *Heihō* in this chapter.

If you master the principles of sword technique, and you reach the level where you can win with ease against one opponent, it is possible to win against any opponent in the world. To win against one opponent is the same thing as winning against thousands or tens of thousands of opponents.

The *Heihō* of a commander is to make major decisions from small bits of information, which is like constructing a great Buddha from a one foot model. It is difficult to write about things such as this in detail. It is a principle of *Heihō* that to know one is to know ten thousand. It is in the *mizu* chapter that such things regarding my school have been written.

Third is the *hi* chapter. This chapter is about combat. From the fact that fires can be large or small, and that the nature of fire is to suddenly and drastically change [which is most striking], it is in this *hi* chapter that large-scale battles are considered. The way of a large-scale battle is the same be it in a one-to-one encounter, or a ten thousand-to-ten thousand encounter. The overall situation should be observed with insight and also with discretion, and carefully examined.

Large things are easy to observe. Small things are difficult to observe. That is to say, it is difficult to actuate one's will with speed with a large number of people; it is difficult to know what is going on with one individual since the spirit of that individual can change very quickly. Matters such as these ought to be carefully considered.

Since the matters which are in this *hi* volume are those which change suddenly, and of situations in which every moment is precious, it is a strength of *Heihō* to be able to fight on as usual because of day in, day out practice. It is for this reason that matters pertaining to winning in combat have been written in this chapter.

The fourth is the *kaze* chapter. This volume is not about

my school, but rather the other schools that exist in the world. I call this the *kaze* chapter because the character *kaze* appears in such words as "old-fashioned," "current," and "family traditions."

Unless one knows well about others, it is not possible to know oneself. There is the possibility of being misled in doing anything. Even if one strives along a path on a daily basis, if the spirit is misguided, it is objectively not the path of truth even if one believes it to be true. Unless one carefully examines in order to ascertain if one's path is the path of truth, the slightest initial distortion later becomes a large distortion. This should be taken to heart.

In the other styles of this world, when one thinks of swordsmanship one envisions "sword techniques." This is quite natural but it is wrong. The philosophy of my style is totally different in its theory and practice from that of other schools. Because of this difference, in order to make known the difference between my school and others in the world, that about the other schools has been written down in the *kaze* volume.

Fifth is the *kū* chapter. The reason why this volume has been titled *kū* is to indicate that there are no secrets or beginnings in *Heihō*. Even if the principles are realized, they must not tie one down. When you are truly free within the path of *Heihō*, an incredible strength surges from within; you react naturally, you become aware of the rhythm of the situation, you face your opponent directly, and take the initiative in attacking. These considerations are the path of *kū*.

KONO ICHIRYŪ NITŌ TO NAZUKERU KOTO
("Why this School Is Named the Two-Sword School")

The *Niten*, or two-sword school of swordsmanship, is so-called because it is the duty of a *bushi*, whether a military commander or a rank-and-file foot soldier, to wear two swords. In ancient times, they were called the long sword, and the sword. Nowadays, they are called the long sword, and the short sword. I do not need to explain in detail why *bushi* wear two swords in this manner. In Japan, be the matter what it may, it is the custom of the *bushi* to wear two swords. My school is called the *Niten* school in order to help people understand the benefits of two swords.

The spear, the halberd, and the like, are weapons for different fights than the sword or the short sword. Even beginners in my school practice with a long sword in one hand and a sword in the other. When in a fight to the death, one wants to employ all one's weapons to the utmost. I must say that to die with one's sword still sheathed is most regrettable.

However, when one holds a thing with both hands, it is difficult to wield it both to the left and to the right; hence my *Niten* school advocates practicing with a long sword in one hand.

The spear, the halberd, and the like, are necessarily big weapons, but both the sword and the short sword are weapons that can be held in one hand.

It is awkward to hold a long sword with both hands when on horseback, running, in a stony field, on a steep road, and in a crowd of people. Because even if one holds a bow, sword, or other weapon in the left hand, one can still wield a long sword in the other. Assuming a stance with a long sword grasped by both hands is not the practical way. When it is difficult to put someone to the sword with one hand, by all means kill him with two hands. It is a matter that requires no great effort. In order to be able to freely employ a long sword in one hand, have people [students] hold two swords and teach them how to wield them.

Anybody, the first time he takes up a long sword in one hand, will find it heavy and difficult to wield.

All things, at first try, are difficult to handle. The bow is difficult to draw and the halberd is difficult to wield. When you grow accustomed to a weapon it gets to be easy to handle. One's strength to draw a bow develops. Also, when one gets used to handling a long sword and learns how to use it, one gains strength and it becomes easy to wield.

As for the way of using the long sword, speed is not the essential aspect. That matter is described in Chapter II, "The Book of Water."

Wielding the long sword in wide spaces, and the short sword in narrow places, is a basic teaching of my school.

In the *Niten* school, one can win with either the long sword or the shorter sword [*katana*]. Accordingly, I do not stipulate the length of the long sword. The spirit to be able to win no matter what the weapon, this is the teaching of my school of *Heihō*.

It is better to wield two long swords rather than just one when facing a number of opponents alone. Also, two long swords are advantageous for taking prisoners. I need not write about such matters in detail here. From one thing, know ten thousand things.

When one learns the principles of *Heihō*, one will be able to understand all things. This must be studied very carefully.

HEIHŌ FUTATSU NO JI NO RI O SHIRU KOTO ("On the Two Characters which Make up Heihō")

In this path, it is commonly said that he who has mastered the use of the long sword is a *heihōsha*.

In the practice of the martial arts, it is said that he who shoots arrows well is called an archer, that he who has mastered firearms is a marksman, that he who uses a spear is a lancer, and that he who is a master of the *naginata*

20

["halberd"] is a halberdier. If that is so, it ought to be that he who has learned the path of the long sword is a "longswordsman," and he who has learned the path of the short sword is a "shortswordsman." Since arrows, firearms, spears, and halberds are all tools of the warrior class, they are all paths of *Heihō*. However, there are specific reasons for calling the path of the *tachi* [long sword], *Heihō*.

Since the virtues of the long sword govern the world and the self, the *tachi* is the foundation of *Heihō*. If one comprehends the virtues of the long sword, it is possible for one to be victorious against ten. If one can win against ten, one hundred can win against one thousand, and one thousand can win against ten thousand. Thus, in the *Heihō* of my school, one opponent and ten thousand opponents are considered the same. All that pertains to what a *bushi* ought to know, not only in the path of the sword, I call *Heihō*.

In speaking of paths, there are paths of the Confucianists, the Buddhists, the men of refinement, the etiquette mentors, and the *Noh* dancers, but these are not in the path of the *bushi*. But even though these are not in the primary path, knowing a variety of paths is very useful. Broaden your knowledge. It is necessary to polish your own path.

HEIHŌ NI BUGU NO RI O SHIRU TO IU KOTO
("Knowing the Advantages of Weapons in Heihō")

When one understands the use of weapons, he can use any weapon in accordance with the time and circumstances.

The short sword is better in confined spaces and when in hand-to-hand combat with an opponent. Generally, the long sword can be used and is good for all occasions. On the battlefield the halberd is in some respects inferior to the bow. The bow can take the offensive, but the halberd goes on the defensive. In the case of an equal degree of skill, the bow is a little stronger than the halberd. According to the circumstances, both the bow and the halberd have few good points in confined spaces. Neither is suitable for taking prisoners. They are strictly weapons for the battlefield. They are essential tools of the battlefield.

Moreover, in learning techniques indoors one is distracted by fine points and forgets the essential teachings of the true path. It will not serve you well in actual encounters. On the battlefield, the bow is useful in tactical advancing and retreating, as it can be fired rapidly from the flanks of the spearmen and others. But it is of little service in sieges of castles or when the enemy is more than 40 yards away.

At present, in many of the martial arts, and especially archery, there is too much concern about appearance, and not much concern about substance. That kind of martial art is useless in critical moments.

From inside a fortress, nothing is better than firearms. Even in field operations, before the start of hand-to-hand combat, they have many good points. But once hand-to-hand combat starts, they are inadequate.

One of the strong points of the bow is that shot arrows can be followed with the eyes. A weak point of firearms is that bullets cannot be seen.

Horses should have great endurance and no habits. As a rule, sturdy and dependable weapons are best. Horses should be sturdy and take long strides. The sword, short sword, bow, and halberd should be full-size and cut well and deep. Bows and firearms also should be sturdy and accurate.

In both weapons and other things, one should not be biased in favor of one thing over another. Too much is the same as too little. Do not imitate or mimic others; one must have a weapon appropriate to one's size and comfortable in one's hand. It is bad for either commanders or rank-and-file soldiers to have likes or dislikes. You must think out these things.

HEIHO NO HYŌSHI NO KOTO
("The Rhythm of Heihō")

There is rhythm in everything, but the rhythm of *Heihō* is something which you cannot gain mastery over without practice.

The rhythms of the path of dance, of the minstrels, and of the wind and string instruments are among the commonly known and obvious ones.

In the path of the martial arts also, there are rhythm and timing for firing arrows, shooting firearms, and riding horses. Rhythm in the various arts also should not be ignored.

There is also rhythm in the abstract. For *bushi*, there is the rhythm of being able to serve, the rhythm of failing, the rhythm of achieving one's purpose, and the rhythm of not achieving one's purpose. In the path of commerce too, there is the rhythm by which one becomes wealthy, and the rhythm by which the wealthy go bankrupt, with the differences in the rhythms according to each path. The rhythm with which things progress and the rhythm with which things deteriorate should be understood and differentiated.

There are many rhythms in *Heihō*. It is one of the main tasks in *Heihō* to first of all learn the rhythm which is appropriate and differentiate it from those rhythms which are inappropriate, to know the differences among the various rhythms for various sizes and speeds in terms of which are appropriate and which are not, and which rhythm will cause the circumstances to be overturned. Your mastery of *Heihō* cannot be considered firm unless you understand the rhythm with which you can avoid being drawn into the rhythm of the opponent.

Victory is achieved in the *Heihō* of conflict by ascertaining the rhythm of each opponent, by attacking with a rhythm not anticipated by the opponent, and by the use of knowledge of the rhythm of the abstract.

Rhythm is a general subject in each chapter. Carefully study what has been written; it is necessary to study this diligently.

Epilogue of the "Earth Book"

The path of *Heihō* of the school which I have so far described, broadens the spirit and is such that it may be called the *Heihō* which encompasses the overall as well as the details. It is in the five chapters of "earth," "water," "fire," "wind," and "emptiness" that this has been written down for the first time.

He who wishes to undertake the study of my *Heihō* should be aware of the following:

First: Do not harbor sinister designs.
Second: Diligently pursue the path of *Niten Ichiryū*.
Third: Cultivate a wide range of interests in the arts.
Fourth: Be knowledgeable in a variety of occupations.
Fifth: Be discreet regarding one's commercial dealings.
Sixth: Nurture the ability to perceive the truth in all matters.
Seventh: Perceive that which cannot be seen with the eye.
Eighth: Do not be negligent, even in trifling matters.
Ninth: Do not engage in useless activity.

The path of *Heihō* should be practiced with the above in mind. In this path it is not possible to become an expert unless truth is pursued with a wide field of vision. One who studies and masters this *Heihō* will not lose even against twenty or even thirty opponents. First of all, if one consistently devotes one's energies to *Heihō*, and aggressively pursues Truth, one should be able to win by the hand and to be obviously superior, and since one's body is completely under one's control due to the training, it is possible to be physically superior, and furthermore, if one disciplines one's spirit sufficiently, it is possible to be psychologically superior. If one can achieve this, how could one lose?

Also, in the broad sense of *Heihō*, it is possible to have good men under you, to use most of these men wisely, to take good care of yourself, to govern the land, to sustain the people, and to keep order throughout. In all of these paths, you can achieve a sense of confidence that you will not lose against others, that strengthens you, and that brings honor to your name. This is the path of *Heihō*.

The twelfth day of the fifth month of the second year of *Shōhō*
[1645]

To Terao Magonojō

Shinmen Musashi

MIZU NO MAKI

("The Book of Water")

Commentary to the "Book of Water" MIZU NO MAKI ("The Book of Water")

> The softest of stuff in the world
> Penetrates quickly the hardest;
> Insubstantial, it enters
> Where no room is.
>
> By this I know the benefit
> Of something done by quiet being;
> In all the world but few can know
> Accomplishment apart from work,
> Instruction when no words are used.
>
> —*The Tao Te Ching*

The *Tao* is a way of life prescribed by the ancient Chinese philosophers and sages. The *Tao Te Ching* is a collection of poems comprising the beliefs of Taoism. Water was used in the *Tao Te Ching* as an analogy for the Way *(Tao)*; both water and the Way permeate everything, with no effort, no sense of doing. *Heihō* is the Way for Musashi, thus the "Water" book. Note also that at the end of this particular poem there are two points made which are also found in *Zen,* and which are part of Musashi's philosophy: the importance of work as a sustaining and fulfilling activity, and the reference to an intuitive learning method.

29

In this book is discussed the psychology and the techniques of Musashi's school of *Heihō*.

● He constantly tells you to always be the same way in any situation, and to keep your mind in the Middle Way attitude.

● The Japanese have a set of concepts labeled *tatemae* and *honne*. *Tatemae* is "what you show to others." *Honne* is "what your real intentions are." Musashi warns you not to reveal your *honne*.

● An *obi* is the belt or sash worn around the outer and/or inner garments. It is tied in a knot. In the manner described, one arranges oneself so that the clothing does not restrict movement.

● *KAN* and *KEN*: *KAN* is seeing through or into, while *KEN* is the observing of superficial appearance. *KAN* is seeing with the mind. *KEN* is seeing with the eyes. The resolution of these two is *Heihō*.

● To observe without moving the eyes: The enhancement of the "sixth sense" was secretly practiced during this period. First the extension of one's peripheral vision was practiced. Sensitivity training was done to heighten awareness of motion and sound, with the intention of producing an intuitive sense of "presence." But here also is the simple concept of awareness, being observant of phenomena as it occurred around you.

● Flexibility is a primary consideration, and not just in the physical sense. The ability to adapt to circumstances psychologically and emotionally was crucial.

● To practice whatever you do the same way all the time is a must. To practice a technique only halfheartedly builds bad habits, and lessens one's practice time of the proper technique. Remember that the context here is life and death swordplay, with razor-sharp, four-foot long lengths of steel. Remember also that dishonesty to oneself was bad discipline.

● The *In-Yo* footwork: When you walk naturally, you walk

with both feet. Even when you simply shift your weight from one foot to the other, both feet will move. This is natural movement. Moving one foot only is not just unnatural, it puts you off balance. The point is to be natural, to move naturally, to think naturally, to move the same way you move every day, even when you are not in combat. To try to change your natural footwork is "mind-stopping."

● Position: Musashi insists that taking a position is not taking a position. When you raise your sword to the upper position, this is not a position from which to move. It is not a position of rest. All movement, all the time, is continuous. It is all part of the final move, that of cutting the opponent. Do not be complacent in a ready position. Always be on the alert. Even though you are protected in this position, always have the same attitude, the same outlook.

● Cutting: Slashing with the sword is called "cutting." Stabbing is used to thrust straight ahead.

● Note that the so-called "secret" of Musashi's school is the *chūdan*, or "middle position."

● The Five *Omote: Omote* refers to the way one faces an opponent, and is called a "play." In modern *Kendō*, the equivalent of swordplay using a "bamboo sword" called a *shinai*, one refers to "playing" *Kendō*.

● Remember that "cutting" is the goal. If you "stop" your mind at blocking or parrying, you cannot cut. Concentrating on the little details of now will cause you to miss the big picture later.

● "The timing of an instant": A *suki* is an interval, literally a space between two objects or in time, where something can enter. This *suki* may be considered as the "stopping" of the mind, a psychological or mental *suki*. It is in this moment that one must strike, when the opponent allows that gap to open. That is why it must be in the timing of an instant.

31

- *Munen musō:* The concept to be gained here is one of spontaneity. Your natural abilities act free from any conscious thought to act. There is no sign of effort; it is an impassive mind. Where there is no intention, there is no thought.
- The long sword instead of the body: For positioning, only the body moves. When cutting, the long sword and the body move as one. The *sword* strikes—*you* do not.
- *Utsu* and *ataru: Utsu* is the conscious dealing of a blow. *Ataru* is to strike without thinking of doing it, that is, just doing it. *Ataru* is the *munen musō* mind at work. *Utsu* would be to put all of your concentration and attention and effort into the blow, which commits one to a course of action from which there is no recovery. *Ataru* is the impassive, "no effort" strike.
- When Musashi talks of the short-armed monkey, which is a variety found in China, he is alluding to a principle of maintaining balance. In both figurative and literal terms, it is the principle of not extending yourself over too great an area. By "pulling away" is meant losing one's balance.
- To stab at the mask: Here we are talking about diversion as a means of attack strategy. Once you have distracted him, gain the advantage by following with your attack.
- Slapping parry: The idea is, in one motion, to both deflect the oncoming sword and to return a strike. The timing involved is a quick one-two movement.

Introduction to the "Book of Water"
MIZU NO MAKI
("Book of Water")

I have entitled this chapter the "Book of Water," because water is the source of inspiration for the method of winning, in the *Heihō* of the *Niten Ichiryū* school. It is beyond my powers to describe this doctrine as I would like to, but no matter how much words fail me, the Truth can probably be understood as self-evident.

Regarding the things recorded in this book, all of them must be considered deeply, word by word. A rough understanding will probably lead to many misinterpretations.

Regarding the way to win in combat, although the events recorded are one-on-one matches, think of them as battles of ten thousand men against ten thousand men. It is essential to view things broadly.

In this doctrine of *Heihō* in particular, once one misperceives the fundamental principles even a little and goes astray, he goes completely off course.

One cannot master the essence of *Heihō* just by reading this book alone. One cannot master the things recorded in this book by just reading the notes and trying to imitate them. They are things that are discovered in a true sense from within oneself. One must exert oneself unceasingly and study very hard.

HEIHŌ KOKORO MOCHI NO KOTO
("The Mental Attitude in Heihō")

The mental attitude in *Heihō* is no different from that of the everyday attitude. In both peaceful times and in times of battle, it is exactly the same. Be careful to ascertain the truth from a broad viewpoint. Do not become tense and do not let yourself go. Keep your mind on the center and do not waver. Calm your mind, and do not cease the firmness for even a second. Always maintain a fluid and flexible, free and open mind.

Even when the body is at rest, do not relax your concentration. When you move rapidly, keep a calm, "cool" head. Do not let the mind be dragged along by the body or the body be dragged along by the mind. Pay attention to the mind and ignore the body. Fully inform the mind and do not be diverted by superfluous matters.

Do not be overly concerned by external matters. Strengthen your fundamental spirit and act in such a way as to not reveal the depths of your spirit to others.

Small men must know thoroughly what it is like to be big men, and big men must know what it is like to be small men. It is essential for both big men and small men to be honest and not be hampered by their physical condition.

One must consider things in general with a clear and open mind. It is essential to diligently improve both knowledge and the spirit.

Broaden your knowledge and know the justice and injustice of the world; know the good and the bad of things. Walk the path of various arts and skills. After you can no longer be deceived by people in general, for the first time, you will attain the essence of the wisdom of *Heihō*.

HEIHŌ NO MINARI NO KOTO
("Posture in Combat")

As for body posture, do not raise or lower the head or lean it to the side. Do not let the eyes wander, and without wrinkling the forehead, form a furrow between the eyebrows. Steady your gaze and try not to blink; narrow your eyes a little more than usual.

With a calm facial expression, hold the bridge of the nose erect and thrust the chin slightly forward. As for the neck, keep the tendon at the back of the neck straight and tense the nape. From the shoulders down, maintain an even distribution of tension throughout the entire body. Lower both shoulders, hold the back straight, and do not stick out the buttocks. Tense the legs between the knees and the toes and tighten the abdomen so the hips do not bend.

There is the teaching which commands us to tighten the knot, and hold the stomach in with the sheath of the short sword in such a manner so as not to loosen the *obi*.

In all the martial arts, it is essential to make the everyday stance the combat stance and the combat stance the everyday stance. You must examine this carefully.

HEIHŌ NO METSUKE TO IU KOTO
("The Point of Concentration in Heihō")

Vigilance in combat means keeping one's eyes wide open. Make *kan* primary ["profound examination of the essence of things"] and *ken* secondary ["observation of the movements of surface phenomena, insignificant actions, what your opponent wants you to see"]. Accurately understanding the state of affairs in the distance and grasping the general situation from the movements near you is most important from the standpoint of *Heihō*. Finding out the swordsmanship [ability] of an opponent and not being deceived in the least by his superficial actions is above all the main object of *Heihō*.

It is necessary to think things out well and to plan well. These rules of vigilance are the same for both the strategy of bouts between two individuals and the strategy of battles involving great numbers.

It is also important to observe both sides without moving the eyes. It is no good trying to learn this kind of thing in great haste. Commit well to memory the things written in this book. Always be watchful in this manner and under no circumstances alter your point of concentration. It is necessary to train hard.

TACHI NO MOCHIYŌ NO KOTO
("How to Hold the Long Sword")

As for the manner of holding the long sword, hold it rather lightly with the thumb and the index finger, neither firmly nor lightly with the middle finger and firmly with the ring and little fingers. It is not good to have slackness in the hand.

One must always take up the long sword with the idea of cutting down the opponent. Also, when one cuts an opponent, grasp the sword without changing your grip, in order to prevent the hand from flinching. Even when you strike, parry or force down an opponent's blade, slightly adjust the position of the thumb and index finger, and in all events grasp the sword with the aim of cutting the opponent.

The grip for a test cutting, and the grip for actual combat, from the standpoint of cutting a man, are no different.

In general, whether in long swords or in the manner of holding a sword, I dislike rigidity. Rigidity means a dead

hand and flexibility means a living hand. One must understand this fully.

ASHITSUKAI NO KOTO
("On Footwork")

As for footwork, slightly raise the tips of the toes and
step somewhat firmly on the heels. In footwork, according
to the circumstances, there are long strides and short
steps, fast and slow . . . but walk with your usual gait. As
for the three steps-called the flying [skipping] foot, the
floating foot, and the firmly locked foot, I dislike all three
of them.

In footwork, *In-Yo [Yin-Yang]* is considered important.
The *In-Yo* foot means not just moving one foot . . . *In-Yo* is
to tread with the feet right and left and left and right,
when cutting, retreating, or parrying. One must not move
just one foot over and over again.

Also, one must take care not to favor one foot over the
other.

GO HŌ NO KAMAE NO KOTO
("The Five Positions")

The five positions, *jōdan* ["upper position"], *chūdan*
["middle position"], *gedan* ["lower position"], *migi no waki*
["right guard position"], and *hidari no waki* ["left guard
position"] are called the Five Directions. Although the
positions are divided into five, they all have the aim to cut
men. As regards positions, there are no others besides
these five.

No matter which position you take, do not think of it as
a position; think only of it as a process of cutting. As for a
great or small posture, it is good to take the most efficacious stance according to the circumstances. The upper,

middle, and lower positions are fixed [firm] positions. The two side positions are fluid. The right and left positions are useful for when there is an obstruction overhead or to one side. Whether to elect the right or left position is to be decided according to the circumstances.

One must understand that the best position, the secret of this school, is the *chūdan* position ["middle position"]. The *chūdan* position is the essence of this school. Figuratively speaking, the *chūdan* position is analogous to the seat of a general in a great battle. The other four positions follow and obey the general. One must study this very hard.

TACHI NO MICHI TO IU KOTO
("On the Way of the Long Sword")

Knowing the way of the long sword means, when one knows the method well, to wield freely even with two fingers the sword that one customarily carries.

If one tries to wield the long sword too quickly, he is mistaking the way, and will become unable to wield it freely. To handle the long sword properly it is essential to handle it calmly.

If one thinks one can wield a long sword rapidly like a folding fan or the shorter sword, he is mistaking the way of the long sword and the long sword becomes difficult to handle. This is called the "short sword cut." One cannot cut a man with this style of using the long sword. When one strikes down with the long sword raise it [straight up] naturally [via the route it descended]. When one strikes sideways, return the long sword sideways. Stretch the elbows out fully and swing the long sword with vigor. This is the way of the long sword.

If one learns well how to use the five basic positions of my school of *Heihō*, one's handling of the long sword will become natural and one will be able to wield the long sword easily. One must train very hard.

ITSUTU NO OMOTE NO SHIDAI, DAI ICHI NO KOTO
("The Five Positions: The First Position")

The first position is the *chūdan*, or *chūdan no kamae*. Point the tip of the long sword at the opponent's face and meet the opponent face-to-face. When the opponent strikes with his long sword, strike it down to the right. Again, when the opponent aims a blow with his long sword return the stroke to the point of his sword and keeping your lowered sword as it is, strike his hand from below when he returns to the attack. This is the "first play" [omote].

These five plays cannot be understood just·by what is written alone. In regard to the five positions, the way of using the long sword must be practiced from the standpoint of the hands. By means of these five positions of swordsmanship, one can know both my doctrine of the way of the sword, and all the sword striking techniques of opponents. Accordingly, I teach that there are no postures other than the five positions of the *Niten* school. One must practice.

OMOTE NO DAI NI NO SHIDAI NO KOTO
("The Second Position")

The second long sword position is "to hold the long sword over the head" [jōdan no kamae] and to strike the opponent the instant he aims a blow at you. Keep the blade that has struck the opponent's sword aside lowered as it is, and when he strikes again, bring the blade up and strike him from below. Should the opponent strike again, repeat the same position.

There are various shades of feeling and timing in this style of fencing. If one practices these techniques by means of my *Niten* school and thoroughly learns the five long sword positions, one can win under all circumstances. One must train diligently.

OMOTE DAI SAN NO SHIDAI NO KOTO
("The Third Position")

As for the third position, "hold the long sword in the lower position" [gedan no kamae], and with a feeling of leading, strike the opponent's hand from below as he attacks. As you strike his hand, the opponent will strike again. Thereupon, as he tries to strike down your blade, strike up and hit the opponent from below. In that fashion one can play the trick of cutting across both the opponent's arms. This is the way of killing the opponent from below, the instant he attacks. The gedan no kamae is practiced both by novice and seasoned swordsmen. In truth, one must train with the long sword.

OMOTE DAI YON NO SHIDAI NO KOTO
("The Fourth Position")

As for the fourth position, assume a stance holding the long sword sideways on one's left side, and strike at the attacking opponent's hand from below. As the enemy strikes down, go along with the flow of his blade, as if to strike the opponent's hand, and cut diagonally up toward your own shoulder. This is the way of the long sword. This is the way, in the case that an opponent should aim a blow at you again, you can win.

OMOTE DAI GO NO SHIDAI NO KOTO
("The Fifth Position")

As for the fifth position, hold the long sword sideways on your right side. Respond to the opponent's attack by swinging your blade up into the upper position cutting diagonally from the lower side and then cut straight down from above. This way of wielding the long sword also is necessary in order to be able to understand thoroughly the

Way of the long sword. When you have become accustomed to wielding a blade in this stance, you will be able to handle even heavy swords freely.

I do not intend to write about the above mentioned five positions in any more detail. Know my way of the long sword in particular and learn timing in general. In order to be able to understand the opponent's long sword, it is first necessary to practice these five positions ceaselessly. While engaging an opponent in combat, also practice this method of swordsmanship. See into the opponent's mind. If you learn various rhythms, you can win by any means.

KAMAE ARITE, KAMAE NASHI NO OSHIE NO KOTO
("The Teaching: Postures and No Postures")

When I speak of having positions and not having positions, I mean that it is not necessary to have fixed long sword positions. However, when it is pointed out that there are five positions, it is indeed possible to have the positions.

No matter which posture one decides to adopt, using the moves of the opponent as opportunities and in response to the circumstances, one grasps the long sword in a manner easy to cut the opponent.

The "upper" [jōdan] position, depending on the situation of the moment, becomes the "middle" [chūdan] position; likewise, the "middle" [chūdan] position, depending on the circumstances, becomes the "upper" [jōdan] position.

Both of the side postures, [holding the long sword extended diagonally at either side of the body] depending on the position, if moved slightly to the center become either the "middle" [chūdan] position or the "lower" [gedan] position.

For these reasons, it is true that there are, and there are

not positions. Above all, when taking up the long sword it is important to cut the opponent. Even if one blocks, strikes, hits, or touches the long sword of the opponent when he attacks, these are all opportunities for cutting the opponent. One must understand this.

When one thinks only of blocking or striking or hitting or holding or touching, one cannot concentrate on cutting.

In all events, it is important to think of all things as a means for cutting. One must study hard.

TEKI O UTSU NI ICHI HYŌSHI NO UCHI NO KOTO ("The Blow of a Single Moment to Hit an Opponent")

The time for striking an opponent is referred to as the strike of a single moment. Take a position within sword's length of the opponent and before he can decide on a move, without moving your body, calmly and spontaneously strike in the timing of a moment.

Timing of a blow before an opponent can decide to retreat, parry, or strike; this is the one timing.

Learn this timing well; you must practice swiftly striking in a split second.

NI NO KOSHI NO HYŌSHI NO KOTO ("The Two-Hip Timing")

This is the kind of situation where one aims a blow at an opponent and he withdraws even faster. In this case, first make as if to strike, then the moment after the opponent flinches for a moment and then relaxes, strike quickly. This is the *ni no koshi no utsu*.

One probably cannot learn to strike just by reading this book alone, but if one receives guidance, one can understand it readily.

MUNEN MUSŌ NO UCHI TO IU KOTO
("On the Blow Free from Worldly Thoughts
—The Spontaneous Blow")

When both you and your opponent decide to strike simultaneously, both (1) adopt the position to strike his body, and (2) concentrate your will to strike. Accelerate your hand naturally and strike hard. This is called the *munen musō* ["free from worldly thoughts or no-mind"] blow, and is a most important stroke. You will encounter it frequently. You must learn it and practice it.

RYŪSUI NO UCHI TO IU KOTO
("On the Flowing Water Blow")

The flowing water blow refers to when one meets with an evenly matched opponent. When the opponent retreats quickly, dodges quickly, and quickly parries your long sword, screw up both your strength and your courage and wield your sword in conformity with this spirit. Very slowly, almost falteringly, like the flow of rivers in the deep parts, strike very deep and very hard. When one masters this stroke, it is very efficacious. On these occasions it is important to determine the opponent's skill, and his strength.

EN NO ATARI TO IU KOTO
("The All-Encompassing Cut")

When one aims a blow and the opponent tries to catch and parry it, with a single movement, strike his head, hands, and legs.

In a single movement, hit him everywhere. This is the *en no atari*.

This strike must be mastered well. It is a style of hitting frequently encountered. One must learn it by experiencing it in detail time and time again.

SEKKA NO ATARI TO IU KOTO
(The "Spark of the Flint" Blow)

By the "Spark of the Flint" is meant to strike with a great deal of force but without raising one's own long sword in the slightest. When your long sword and that of the opponent are close enough to be barely touching, strike with a great deal of force but without raising your own long sword in the slightest. It is necessary to strike quickly. It is not possible to deal this blow without practicing it over and over. If practiced sufficiently, it is possible to strike with force.

MOMIJI NO UCHI TO IU KOTO
("The Scarlet Maple Leaf Blow")

The scarlet maple leaf blow means to strike down the opponent's long sword with the intent to gain control of it.

When an opponent with a long sword assumes a position in front of you with the intent to strike, parry his long sword with *munen musō* ["impassive state of mind"] strikes, or *sekka* strikes. If the strikes are dealt out in such a way as to lower the sword point of the opponent with purpose, it is certain that the opponent will drop his long sword.

TACHI NI KAWARU MI TO IU KOTO
("The Long Sword Instead of the Body")

"The long sword instead of the body" could also be called "the body instead of the long sword." Whenever one strikes an opponent, the long sword and the body are not moved simultaneously. First, the body assumes a striking stance, [the stance] being dependent upon the position from which the opponent is striking, and then the long sword follows up a little later to strike.

There are instances when the long sword strikes while the body is immobile, but usually the body moves first, and after it the sword strikes. This should be carefully studied and practiced.

UTSU TO ATARU TO IU KOTO
(On "Utsu" and "Ataru")

To *utsu* and to *ataru* are separate. To *utsu* is "to consciously deal a blow, regardless of the way in which it is dealt." To *ataru* has the meaning of "to come by" regardless of how strongly one does this so that it is to *utsu* even if it be strong enough so as to instantaneously kill the opponent. To *utsu* is to do so consciously. This should be researched.

To *ataru* against the hands or the legs of an opponent is first of all to *ataru* and this is done so as to be able to *utsu* afterward. To *ataru* means to come into contact with one's opponent. Practice in order to be able to differentiate between these two. One should learn to improvise.

SHUKŌ NO MI TO IU KOTO
("The Body of the Short-Armed Monkey")

By the "body of the short-armed monkey" is meant the idea of not extending one's hands. It is the trick or technique of quickly leaning toward an opponent just when he is about to strike, where you bring your body close to that of the opponent without sticking out your hands. If one sticks out one's hands there is always a tendency for your body to be pulled away. Therefore move your whole body quickly closer into the opponent. This should be studied carefully.

45

SHIKKO NO MI TO IU KOTO
("The Body of Lacquer and Glue")

By "lacquer and glue" is meant to stick closely [in the sense of remaining in contact] to the body of one's opponent and not to become separated from him. When one approaches the body of an opponent, stick your head, body, and legs very close to him.

Most people, even if they stick their legs and heads closely, keep their bodies away. Keep your body also close to that of the opponent so that there is no space in between. This should be considered carefully.

TAKEKURABE TO IU KOTO
("Comparing the Height of Bamboo")

By "comparing the height of bamboo" is meant that whenever you keep your body close to the body of the opponent, do not allow your body to "scrunch up," but rather, stretch your legs, torso, and neck as well and keep close to the opponent so that one is face-to-face with him as if one could win by being the taller one. One must be resourceful about this.

NEBARI [Tenacity] O KAKURU TO IU KOTO
("Stick-to-it-iveness")

When both you and your opponent attack with long swords, and the opponent parries your attack, with great determination, keep your long sword firmly against his.

To have tenacity means that your long sword cannot readily be pushed away, and not that you approach with great force. When you are approaching while keeping your long sword against that of the opponent, it does not matter how quietly you bring in your body.

There is stick-to-it-iveness and there is getting entangled;

stick-to-it-iveness is strength and entanglement is weakness. You must know the difference.

MI NO ATARI TO IU KOTO
("On the Body Strike")

By *mi no atari* [body strike] is meant to push one's way into the space of the opponent, and to hit into the body of the opponent. Turning one's face slightly to the side, thrust out one's left shoulder and strike the chest of the opponent.

Use as much strength as possible when closing, with your breathing under control, with firmness, so as to cause the opponent to bounce off. Resolutely charge into the chest of the opponent.

If you continue to practice this charging in, it should get to the point that the opponent will be thrown eight, sometimes even twelve feet, even at times to the point of killing the opponent.

MITSU NO UKE NO KOTO
("The Three Ways to Parry")

By the "three ways to parry" is meant that when the opponent comes charging into you that you ought to, in order to parry the opponent's long sword, hold your long sword in such a way so as to stab the eyes of the opponent, making his long sword go off to your right side.

Or, when parrying a lunge, get the long sword of the opponent to stab the right eye of the opponent so that the lunge is parried in such a way as to sandwich the neck.

Also, when the opponent is striking and when you are going in with a short sword, do not be too concerned about your parrying long sword, and with the left hand go in as if to stab the opponent's face.

Although these are the three parries, think as if your

left hand is a fist with which you will punch the opponent's face.

OMOTE O SASU TO IU KOTO
("To Stab at the Face")

By "to stab at the face" is meant that when the long swords of the opponents and allies are equal, to stab the face of the opponent with the point of one's own long sword between the long swords of the opponent's and the long swords of one's own side. If there is the will to stab the face of the opponent, the opponent will endeavor to move away his face and torso. If the opponent tries to move away his face and torso, there are many ways by which one can achieve victory. This should be thought out.

During combat, if the opponent intends to get out of the way, for all intents and purposes you have already won. That is why this "to stab at the face" should not be forgotten. While practicing the martial arts, one should train well in this advantageous method.

KOKORO O SASU TO IU KOTO
("To Stab the Heart")

By "to stab the heart" is meant, when engaged in combat, in those locations when that which is overhead and that which is on the side are jammed so that there just is not any way to cut, the opponent should be stabbed.

In order to avoid the opponent's long sword, show the opponent the back of your sword, vertically, and draw it in such a way that the tip of the long sword does not waver, and stab at the chest of the opponent. When you are tired or when your sword does not cut well, use this method. The decision to use this method should be made carefully.

KATSU TO IU KOTO
(On "Calls")

Both *katsu* and *totsu* calls are used when you are attacking and when you are pinning down the opponent. Raise your sword so as to stab from below where the opponent is likely to attempt to return the blow, thus striking with the return strike. Both moves are made with a quick, decisive rhythm, with the strike being made with the *totsu* and *katsu* calls, and with the *katsu* call being made with the upward stab, and the *totsu* call being made with the striking move. This rhythm is that which is usually encountered in a match. This *katsu-totsu* method is made with the intention to stab the opponent with an upward motion of the cutting end, and it is a rhythm which is used at the same time, with decisiveness, that the sword is raised. This must be carefully practiced and studied.

HARI UKE TO IU KOTO
(The "Slapping Away" Parry)

In combat, if the rhythm has been disrupted and cannot be regained, when the opponent strikes you, beat back the attack with the long sword by striking.

To slap away is not done so strongly, nor is it parried so strongly. Depending on the long sword with which the opponent is striking, slap away his sword, and the important thing is to strike the opponent more quickly than you would slap him away. It is important to go on the offensive by slapping him away and striking.

As one becomes an expert in slapping away, no matter how strongly the opponent strikes, as long as there is the intention to slap him away even slightly, there is no chance that the point of your long sword will be dropped. This ought to be well practiced and studied.

TATEKI NO KURAI NO KOTO
("The Order of Opponents when Fighting Alone")

The *tateki no kurai* is meant for those occasions when one person fights a large number of opponents.

Pull out both the long sword and the short sword and assume a position of wielding the long sword horizontally, swinging in large sweeps from left to right and back. Even if the opponents attack from all four sides, one fights by pushing them away in one direction.

When the opponents come forward in attack, judge carefully which of the opponents are coming in first and which ones are coming in later.

Paying heed to the overall movements of the opponents, ascertain the rhythm in which the opponents come in to strike, and fight in such a way that the sword on the right and the sword on the left cross at once. Do not wait. Immediately put both arms in a position of readiness and cut in strongly where the opponents come in, so as to push and crush them, striking as the opponents come forward with the intention of scattering and crushing them.

No matter what you do, assume that you will beat back the opponents as if they were one line of fish which are connected to each other, and when you have seen that the ranks of the opponents have been disarrayed and that they are getting on top of one another, push in and strike strongly without allowing any time to lapse.

Merely pushing back wherever the opponents are clumped together is of no use. Also, merely striking opponents as they come forward leads one into a frame of mind of just waiting, and this is useless. Study well the rhythm with which the opponents strike and learn how to go about breaking it up.

Whenever possible, gather a large number together and

practice pushing forward by pressing your body forward so that one on ten or even on twenty becomes easy. This should be very carefully practiced and studied.

UCHIAI NO RI NO KOTO
("Principles of Exchanging Blows")

This is to know the path by which one achieves victory with the long sword of *Heihō*. I will not describe this in detail. By diligently practicing this, one ought to be able to learn the ways of winning.

Use the long sword, which ought to be called the Way to the essence of *Heihō*. This is left to oral tradition.

HITOTSU NO UCHI TO IU KOTO
("The One Strike")

By means of the "One Strike" it is possible to be sure of achieving victory. Unless one thoroughly studies *Heihō*, it is not possible to understand this. If these principles are well practiced, they can be used at will, and it will be possible to obtain victory as you please. This will not be possible unless it is well practiced.

JIKITSŪ NO KURAI TO IU KOTO
("The Meaning of the Spirit of Direct Communication")

The "spirit of direct communication" is to learn the true path of *Niten Ichiryū* and to pass it on. It is important to practice diligently and to make this *Heihō* a part of yourself. This is oral tradition.

Epilogue of the "Water Book"

What I have written down in this chapter is the essence of the sword techniques of my school.

In order to master the ways by which one can be victorious over others with the long sword, first, one must learn the five positions by means of the five *omote* [external appearances]. Once this occurs, you will learn the path of the long sword, your entire body will move at your will, and you will know the rhythm of the path with your spirit. Finally, your self, long sword, and hands will be skilled. The movements of your body and legs will be in coordination with your spirit. Consequently, one will gradually attain the principles of this path, winning against one opponent, and then two opponents, and then become capable of distinguishing between the good and the evil in terms of *Heihō*. This will be done by practicing what is taught in each and every passage of this text. You must walk down the path of a thousand miles step by step, keeping at heart the spirit which one gains from repeated practice with whomever one can get to practice with, and knowledge attained from whatever experiences you can come by, without impatience.

Consider the following with patience: It is the role of the *bushi* to undertake this *Heihō*, to win tomorrow against someone who is lesser than you, and then to win against someone who is greater than you, in a manner which follows this text precisely, and which does not allow the spirit to deviate in the slightest, without allowing in the least, one's mind to wander off.

No matter how much one manages to win against opponents, those victories are not in the true path, if they are accomplished by means which are contrary to the teachings of this school. If the heart of this teaching is a part of you, then you should have an understanding of how to win against several tens of opponents by yourself. With the strength of the knowledge of swordsmanship one should master *Heihō* of large armies and the *Heihō* of individuals.

Practicing a thousand days is said to be discipline, and practicing ten thousand days is said to be refining. This should be carefully studied.

The twelfth day of the fifth month of the second year of *Shōhō*
[1645]

To Terao Magonojō

Shinmen Musashi

HI NO MAKI

("The Fire Book")

Commentary to the "Fire Book" HI NO MAKI ("The Fire Book")

In the Fire Book, Musashi discusses strategy. He discusses the way to confuse an opponent, when to attack, and what to think about when routing the enemy. Most of the teachings here need no explanation. Just imagine that you are on the battlefield, and follow his instructions. Your battlefield can be any endeavor where you must deal with people.

In the introduction to this chapter, understand what is meant by supernatural powers. It is not that Musashi is claiming that his *Heihō* will provide you with such powers. Rather, he means that with such utter confidence in your abilities, no worldly power will be able to win over you. Although it may seem as if boasting, it is not. It is the knowledge that his *Heihō* works, from experience, that gives him the right to say it. He is stating facts. And he is not alone. A little known contemporary of Musashi, Odagiri Ichiun, a *Zen* master and swordsman, makes an identical claim about his own situation. "I am the only swordsman with no peers in the world." A battle between these two would have been interesting. Nevertheless, understand it as an affirmation of confidence.

● Know the rhythms of the intervals: Discussed previously is the concept of *suki*, the gap in activity, or consciousness, that one can take advantage of. When the opponent tenses, or slackens; if his eyes blink or if he takes a breath, these are *suki*.

● To injure the corners: It is a principle of *Jūdō* to direct an attack to the opponent's "balance points," which are the four corners in which he is vulnerable. In any martial art, this is a guiding principle.

Introduction to the "Fire Book" HI NO MAKI ("The Fire Book")

In the *Heihō* of my *Nitō Ichiryū* school, combat has been likened to fire. Those matters pertaining to victory or defeat have been written in this "fire" chapter.

First of all, most people tend to consider the principles of *Heihō* on a small scale when they make their interpretations of it. They strive to win by movements of the fingertips, or movements of the wrist by a few inches. They believe that victory can be decided by the speed of the movements of the forearms as if they were waving a fan, or think that if they are just a bit faster than their opponent with the *shinai* [bamboo sword used for practice], that they have an advantage. They devote themselves to the practice of arm and leg movements in order to be as swift as possible.

In my *Heihō*, which has withstood exchanges in combat time and again, where my life was at stake, where I was faced with life-and-death situations, I have learned the Way of the Sword. I have experienced the strengths and weaknesses of the opponent's striking long sword; I have come to know the way of using the cutting edge and the back ridge of the sword; I have been disciplined by the experience of striking down opponents. Therefore, it does not occur to me to think about such minor details and insignificant matters. Particularly, when one is engaged in battle in full armor, these techniques become even more insignificant.

Furthermore, in a contest where you are fighting for your life, it is the path of my *Heihō* to know the way to be victorious, one alone against five or even against ten. Accordingly, how could there be any difference between the principles of winning one against ten and a thousand against ten thousand?

Nevertheless, it is not possible in day-to-day practice to bring together a thousand or tens of thousands of people to practice this path. Even when one is practicing alone with a long sword, you can attain an understanding with which to win against ten thousand, by virtue of this *Heihō;* you can attain proficiency in this path by seeing through the opponent's strategy, and by knowing the strengths and weaknesses of the opponent's tactics.

If you think that you are the only one who can attain my *Heihō* in the entire world, if you are deeply committed to the eventual mastery of this path, if you practice day and night polishing your skills through and through, then you will be the only one who can attain such freedom and such power to perform miracles. You will attain supernatural powers. This is the secret of *Heihō*.

BA NO SHIDAI TO IU KOTO
("Considering the Site")

Establishing the site of conflict so that it is advantageous is very important. The primary principal is to have the sun behind you. Take a stance with the sun behind you. If it is not possible to have the sun behind you because of the lay of the land, keep the sun to your right.

In a room also, it is the same in that the light should be behind you or to your right. Thus, it is desirable to have the space behind you not open to use, with plenty of room to your left, and to push forward and to protect the right side. At night also, in a place where you can see the opponent, you should, as in the above situations, endeavor to take a stance such that the fire is behind you and the light is to your right.

One must strive to take a stance in which one is as high as possible so as to look down on the opponent. [Japanese etiquette prescribes that the highest seat is the seat of honor.]

Now, in actual battle, one should endeavor to chase the opponent around by going to the left, but it is of the utmost importance to force the opponent into a disadvantageous position. Once cornered, the opponent should be chased so that he does not have the opportunity to look around, that is, do not let him see where he is. The same applies in a room; chase the opponent into the doorsill, lintel, doorway, sliding panels [shōji], balcony, or pillars so that he has no opportunity to look around.

In any and all situations, when chasing the opponent, it is important to be in an advantageous position by taking advantage of any places that afford difficult footing, that have obstacles on the sides, or other such characteristics. This should be carefully and thoroughly studied and practiced.

MITSU NO SEN TO IU KOTO
("What is Meant by the Three Initial Attacks")

The first of the "three initial attacks" is the *Ken no Sen*, where you make the initial move. The second, the *Tai no Sen* ["The Waiting Attack"], is where your initial move takes place instantly after the opponent makes the first move. The third one is the *Taitai no Sen* ["The Body and Body Attack"], which is the initial attack made when you and the opponent attack at the same time.

In any conflict there can be no other initial attacks other than these three. Since it is possible to achieve a quick victory depending on the way in which the initial attack is made, this concept of "first move" is of primary importance in *Heihō*.

Although there are numerous aspects to this concept of "first move," it is not such that should be written out in detail, since the question of which "initial attack" ought to be taken is primarily answered by the circumstances of each situation. Victory is achieved by reading into the designs of the opponent and the knowledge of my *Heihō*.

● First, the *Ken no Sen* ["The First Attack"]. This is when one considers taking the initiative by attacking first. You quietly assume the position from which to initiate the attack and then swiftly make the attack with no hesitation. This should be an initial attack which on the surface is very forceful and fast, but which leaves you some reserve. Do not spend all your energy on your first attack. Also, this is an attack which is made with substantial strength of will, with the leg movements made faster than usual so that when the attack is made, one closes in on the opponent in one swift breath. Also this is an initial attack in which one empties out one's spirit. From beginning to end, you thoroughly overpower the opponent with enthusiasm and forcibly win under any circumstances. This is under the category of "*Ken no Sen*" [The First Attack].

● Second, the *Tai no Sen* ["The Waiting Attack"]. First of all, when the opponent lunges forward toward you, make as if it does not bother you in the least and feign weakness. When the opponent has come in quite close, suddenly increase the distance by backing and make it seem as if you are leaping away. Come in forcibly in one short breath and win as the opponent shows signs of slacking; this is one of the ways of "*Tai no Sen*" [The Waiting Attack]. Also, when the opponent attacks with force, if you counterattack with an even more forceful attack, the rhythm with which the opponent attacks is altered. Take advantage of that moment of change, and attain victory. Such are the principles of *"Tai no Sen."*

● The third, *Taitai no Sen* ["The Body and Body Attack"]. This is the attack which is used when the opponent comes forward quickly, and one responds quietly with strength, and when the opponent has come in sufficiently, one suddenly attacks in one swift breath and achieves victory. When the opponent comes in quietly, you move yourself a little more quickly than usual in a manner which can be likened to floating, and at the point when the opponent has come in close enough make a feinting move, watch the reaction of the opponent, and then quickly attack to win. The above is the *"Taitai no Sen."* It would be difficult to write and explain more as to the details of this.

What has been written in this section is merely the basics and should be elaborated upon. Regarding these three *Sen*, it should not be taken that one must always be the first one to attack regardless of the situation or circumstances, but at the same time, it is generally desirable to be the one to initiate the attack and thereby put the opponent in the defensive position. In either case, whether one is attacking first or the opponent initiates the attack, the concept of initial attack is a principle of the knowledge of *Heihō* with which one always achieves victory.

MAKURA O OSAYURU TO IU KOTO
("To Restrain the Pillow")

To "restrain the pillow" means not allowing the head to be raised. It is especially bad to be dragged around by the other side or to be placed on the defensive, particularly in the path of *Heihō*. No matter what happens, the ideal is to be in a position in which one can freely lead the opponent around, that is, to be on the offensive.

It can be assumed that the opponent will think likewise,

and that you will think likewise, but all this is of no use unless one is aware of how the other side will come forward. To "restrain the pillow" as it is understood in *Heihō*, is to stop the opponent as he tries to strike, to restrain the opponent as he tries to lunge, and to wrench away as the opponent tries to grapple with you. This means that he who is well versed in my *Heihō*, when crossing swords with an opponent, and regardless of what the opponent does, will know in advance the designs of the opponent. When the opponent tries to strike, the opponent will be stopped at the very onset of his attempt, at the *S* of strike, and not be allowed to continue on. For example, you must stop the opponent when he tries to attack at the *A* of the attack; you must stop the opponent when he tries to jump at you at the *J* of the jump; you must stop the opponent when he tries to cut you at the *C* of the attempt to cut.

When the opponent sets up a move, it is important to leave that which is of no use to the opponent and to hold down that which can be of use so as to make it impossible for the opponent to carry out his plans. To always be trying to hold down the opponent is, in itself, to be placed in the defensive. He who is skilled in *Heihō*, first of all, accomplishes all of his moves, no matter what, according to the Way. When the opponent attempts to execute a move, frustrate it from the onset, make whatever the opponent was trying to accomplish of no use, and achieve the freedom with which to lead the opponent. This too is the result of practice. To "restrain the pillow" should be well and carefully appreciated.

TO O KOSU TO IU KOTO
(What is Meant by "to Cross the Expanse")

When one speaks of "crossing the expanse" it can be in the context of crossing a sea or crossing a channel. It can be a short distance or a long distance. In the course of a lifetime there are usually a number of difficult situations which could be likened to crossing an expanse. The "expanse" is crossed by piloting the boat, by researching the location of the "expanse" if it is located on a sea route, by knowing the performance capabilities of the boat, by knowing well the favorable and the unfavorable points regarding the weather conditions, by making the necessary adjustments according to the conditions, regardless of whether another boat or other boats will be accompanying your boat, by relying on a crosswind or by being pushed by a tail wind, and if the wind direction changes, by rowing for three to five miles, all with the intention of reaching the port.

In order to pass through life, there is the need to have a spirit, to be decisive about exerting all of one's energies to overcome difficulties.

In *Heihō* and in battle, also, crossing the expanse is important. Overcoming a difficulty, knowing the extent of the opponent, and being aware of one's strengths and acting according to the principles of *Heihō* is the same as an excellent captain crossing a sea route.

Once you have overcome the difficulty, you can feel safe. By crossing the expanse, the weaknesses of the opponent come to light, and one is placed in a position of superiority so that in most cases, it is possible to achieve victory. Crossing the expanse is important; be it in terms of *Heihō* as applied to a conflict which involves many, or conflict which is one-on-one. This ought to be well and carefully appreciated.

KEIKI O SHIRU TO IU KOTO
(What it Means to "Know the Prevailing Conditions")

To "know the prevailing conditions" means to make the decisions as how best to move one's own troops and which strategy to use in conflicts involving a large number of people; knowing whether or not the spirits of the opponents are high or are waning, knowing the psychology of the opponent's troops, having a grasp of the prevailing conditions of the site of conflict, and observing the conditions of the opponents.

In one-on-one conflicts it is essential to understand the flow of the opponent's personality, to find out his strengths and weaknesses, and to plot against the opponent's expectations. Know the ups and downs of the opponent. Know well the rhythm of the intervals between them, and thereby take the initiative.

The flow of things can always be seen if one's intelligence is good.

Once *Heihō* becomes a part of you, you will be able to surmise the opponent's thoughts, and to think of numerous ways in which to achieve victory.

This must be fully thought out.

KEN O FUMU TO IU KOTO
(What is Meant by "to Tread on the Sword")

"To tread on the sword" is something which is commonly used in *Heihō*. First of all, in *Heihō*, as it pertains to large numbers, even when bows and firearms are involved, when the opponents attack you, the opponents will probably come in to attack after they have shot their bows and firearms at you, so in response [to this type of attack] if you [waste your time] preparing bows or loading your firearms, you will not be able to charge in when you should make your attack. It is important to make a swift attack while the opponents are firing their bows and firearms. By attacking quickly the opponents will be rendered incapable of utilizing their arrows. They will also not be able to fire their firearms. This means to parry the attack as it comes from the opponents and win

66

with a spirit to tread over what the opponent is doing.

In *Heihō* as it pertains to one-on-one also, if you strike after the long sword of the opponent strikes, the rhythm will be something like *to'tan*, *to'tan* [a light beat, the "*to*," followed by a harder "thump" like beat, the "*tan*"], and matters will not progress. You must make it impossible for the opponent to attack a second time by striking at the place from where the opponent is about to strike out with the spirit of treading on the long sword with which the opponent strikes. To tread is not a concept which is limited to the feet. One must have the intent to tread with one's body, spirit, and of course with one's long sword so as to render the opponent incapable of a second round.

Thus, this is the spirit which strives to take the initiative. Although it has been said that one ought to have one and only one encounter with the opponent, this does not mean that one should crash in, but rather, it means that one should have the intent to remain holding him down. This should be carefully investigated.

KUZURE O SHIRU TO IU KOTO
(What is Meant by "to Know Collapse")

"To know collapse" is something which pertains to all things. Houses collapse, the body collapses, and the opponent collapses, all when their time comes and their rhythm is broken.

In *Heihō* as it pertains to large numbers, it is important to chase the opponent so as not to lose the moment afforded by seizing the rhythm of the opponent's collapse. If you lose the moment afforded by the opponent's collapse, the opponent may recover.

In *Heihō* as it pertains to one-on-one also, there are usually times during the conflict when the rhythm of the opponent goes haywire and he begins to collapse. If you miss that by not being vigilant, it is possible that the opponent will recover and that a stalemate will develop. It is important to keep an eye out for such a collapse and to strike and chase with certainty so that the opponent does not have the opportunity to recover. To strike and chase is to strike with force; in one breath deliver a strike which will render the opponent incapable of recovery. There is the need to understand well the delivery of this strike. If the delivery is poor, it will leave much to be desired. Improvise on this.

TEKI NI NARU TO IU KOTO
("To Become the Enemy")

"To become the enemy" is to think as if one were the enemy. Look around you and you will see that even the thief who blockades himself in a house [when caught in the act] is considered by his opponents to be most formidable. But if you put yourself in his position, [you can see that] he feels helpless, that everyone in the world is against him. He who has blockaded himself is like a pheasant, while he who is waiting outside is like a hawk. There is a need to consider this carefully.

In *Heihō* as it pertains to large numbers, one [tends to] think(s) of the opponents as strong and becomes passive. But there is no reason to fret if one has good troops, understands the principles of *Heihō* well, and comprehends well how to win against the opponents.

In *Heihō* as it pertains to one-on-one also, you ought to put yourself in the opponent's position. If you believe that you are up against a swordsman who is proficient in the Way, you will lose. Examine this closely.

YOTSUTE O HANASU TO IU KOTO
("To Release the Four Hands")

"To release the four hands" is to know when to give up whatever you were aiming to do, and to win by another means when both you and the opponent are of the same spirit and you have gotten into a state of being locked with each other so that no progress can be made.

In *Heihō* as it pertains to large numbers, if there is a "four-hands" impasse and a locked-in situation, it is not possible to reach a decision and the casualties will be heavy. The first thing to do is to quickly discard the previous designs and come up with something which the opponents do not anticipate.

In *Heihō* as it pertains to one-on-one also, it is necessary to change your designs, ascertain the opponents' condition, and achieve victory by other useful means. Learn to judge this well.

KAGE O UGOKASU TO IU KOTO
("To Move the Shadow")

"To move the shadow" refers to a method for the occasions when you cannot determine the intentions of the enemy.

In large-scale battles, when no matter what you do you are unable to determine the enemy's situation, pretend that you are going to attack very fiercely, and you will learn the enemy's plan. When you know the enemy's plans, it will be easy to gain victory by means of an appropriate response.

Again, in single combat also, if the opponent adopts a stance with the long sword to the rear or to the side, and you cannot guess his intentions, if you try, make so as to strike; the opponent will reveal his intention with his long sword. When you know the opponent's plan, you can most certainly gain victory by means of an appropriate response to it. However, if you are careless, you can completely lose the timing of your blow. You must study diligently.

KAGE O OSAYURU TO IU KOTO
("To Suppress the Shadow")

"Suppressing shadows" refers to the method adopted when you can determine the morale of the opposing side.

In large-scale battles, when the enemy starts to employ a particular tactic, if you show him your determination to check him, your enthusiasm will overwhelm him and he will change his tactics. Thereupon, you also change your tactics, seize the initiative, and gain the victory.

In single combat also, check the enthusiasm of the opponent by means of your timing. In this manner, find the way to seize the initiative and win. You must plan diligently.

UTSURAKASU TO IU KOTO
(What is Meant by "to Make Transferable")

The concept of "to make transferable" exists in all things. Sleepiness and yawning can be transferred. Time, too, can be transferred.

In conflicts which involve a large number of people, it is possible to get the opponents to become lax in their guard. When they are in a state of agitation and show signs of impatience, appear as if nothing is bothering you and put forth an easygoing, relaxed stance. When you perceive that the mood has been transferred, you have a chance to achieve victory by making a strong attack with as much speed as possible.

In one-to-one conflicts also, it is important to let the opponent see you relaxed in both body and spirit, and to achieve victory by means of an offensive move made with strength and speed.

There is also the concept of "intoxication," which is similar. It involves getting the opponent bored, agitated, and giving him the feeling that you are inferior. This should be well thought out.

MUKATSUKASURU TO IU KOTO
(What is Meant by "to Upset")

In all situations it is possible "to upset," or unbalance, and to anger the other side. One way is to make him feel endangered; the second is to make him feel that it is impossible; the third is unforeseen situations. There is the need to research this well.

Even in conflicts which involve a large number of people, it is important to upset and enrage the other side. It is important to win by means of an advantageous move, before your opponent has the chance to calm down his spirit after having been on the receiving end of an attack which has been made with ferocious fervor at a time when the opponent least expected such an attack.

Also, in one-to-one conflicts, it is important to achieve victory by first confronting the opponent with a slow-paced approach, and then suddenly making a forcible attack. Take advantage of the agitated state of the opponent, without giving the opponent even a moment to catch his breath. This should be researched well and carefully appreciated.

OBIYAKASU TO IU KOTO
("On Threats")

Fear of things is not rare. One is often filled with feelings of dread by the unknown.

In large-scale battles, it is not visible things alone that

induce fear in the enemy. One can probably frighten him with noise, or by making a small force appear to be a large one, or by attacking suddenly from the sides. These are all ways of inducing fear into one's opponents. One can win by taking advantage of the enemy's confusion, and loss of rhythm.

In single combat too, one should induce fear into the opponent with one's body, long sword, and voice. It is important to suddenly do something that the opponent does not expect and to take advantage of his fear to defeat him. One must study this very hard.

MABURURU TO IU KOTO
("To Soak In")

"Soaking in" is the important tactic employed when you and the opponent strenuously engage one another, and you decide that you cannot come to a conclusion; leave things as they are, "soak in" to the opponent and while there, employ an advantageous method to gain victory.

In both great battles and small engagements, when enemies and allies are separated and facing each other, and contending but cannot come to a conclusion, it is important to win in a flash by closing with the enemy, and engaging him in hand-to-hand combat. Under these conditions, discover the road to victory by advantageous means. One must study very diligently.

KADO NI SAWARU TO IU KOTO
("To Injure the Corners")

This refers to the fact that when attacking a strong force, it is difficult to attack it directly as it stands. In these cases, one attacks the corners.

In large-scale battles, after a careful inspection of the enemy's forces, one can gain advantage by attacking the

corners of exposed strategic points. W..........
nated the strength of the corners, the...........
whole also will be diminished.

Even as that strength is being eliminated,...........
tant to strike the strategic points and gain victo.....

In single combat also, if one injures the corne.......
opponent's body, the body will gradually weaken, and
when it reaches the verge of collapse, winning is easy.
Study this well. It is important to understand the tricks for
winning.

UROMEKASU TO IU KOTO
("Confusion")

"Confusion" refers to denying any feelings of certainty
in the mind of the enemy.

In large-scale battles, determine the enemy's aims on
the field of battle and by means of the mental power of my
Heihō confuse his mind in various ways as to whether you
intend to move here or there, this way or that way, fast or
slow. Employ timing to induce a flurried mood in the
enemy; this certainly is the way to understand the principles of winning.

In single combat also, seize opportunities, and employ
various tricks: Feint blows and thrusts and make your
opponent think that you are going to strike him. When, in
this manner, you have induced a state of confusion in the
opponent, take advantage of it and win in accord with your
plan. This is the spirit of combat. One must study this
well.

MITSU NO KOE TO IU KOTO
("Three Yells")

The "three yells" are divided into the *shō*, or pre-yell,
the *chū*, or during-yell, and the *go* or post-yell. Depending

...nces, yelling is very important. Because ...ourages us, we yell at such things as fires, and ...he wind and the waves. Yells show spirit.

...n large-scale battles, the yell given at the outset of combat is loud in order to overawe the other side. Again, yells given during combat are pitched low from deep within the abdomen. Furthermore, following victory in battle, the yell is strong and loud.

In single combat also, one yells "*ei*" just before initiating a strike in order to shake up the opponent, and after the yell, delivers a blow with the long sword. Again, the yell given after scoring a hit on an opponent is the yell of victory. These two are called the *sengo no koe,* or the "before and after yells."

One does not yell simultaneously with the delivery of a strike with the long sword. Again, because the yells are used in actual combat to help maintain timing, they are pitched low. These are things to study diligently.

MAGIRURU TO IU KOTO
("To Mix Up")

"To mix up" means, in the case of a large-scale battle where bodies of troops confront each other and it is observed that the enemy is powerful, to attack one side of the opposing body of troops and upon seeing it put into disorder, to immediately withdraw and to attack again at another strong point. In short, it means to fight in a zigzag fashion.

This is also an important trick when facing a large number of opponents alone. When one defeats one individual or forces him into retreat, attack again at another strong point. As you learn the enemy's situation, attack with good timing to the left and right in a zigzag manner.

Ascertaining the opponents' degree of skill and attacking

74

with a determination not to pull back is a forcible way to win.

This spirit is also necessary in single combat when closing with a strong opponent. "To mix up" refers to the trick of closing with the enemy with the determination not to retreat a single step. It is essential to fully appreciate this.

HISHIGU TO IU KOTO
("To Crush")

"To crush" refers to the trick of regarding the opponent as weak and thinking of yourself as strong, and crushing him at a stroke.

In large-scale battles also, when one sees that the enemy is few in number, or even if many, if the enemy is confused and fainthearted, immediately concentrate your strength and crush him completely. If one deals with him in this manner too lightly, he can recover. It is important to understand fully the knack of grasping things in the fist and crushing them completely.

Again, in the case of single combat, when the opponent is less experienced than you, or his timing is flurried and he seems to be on the verge of fleeing, one must crush him at a stroke without giving him pause to take a breath or to exchange glances. Most important, you must allow him absolutely no chance to recover himself. It is essential to study diligently.

SANKAI NO KAWARI TO IU KOTO
("Mountain and Sea Change")

The spirit of "mountain and sea change" refers to the fact that while one is engaged in combat with an opponent, it is bad to frequently repeat the same tactic.

Repeating the same tactic twice is sometimes unavoidable, but it is not to be done three times. If in the process of performing a trick on the opponent, you fail once, even if you try again, you will meet with no more success than the first time.

Above all, try a different approach and take the opponent by surprise. If it fails, you must try something different again.

In this manner, if the opponent expects mountains, give him the sea; if he expects the sea, give him mountains. Taking people by surprise is a teaching of *Heihō*. This must be studied diligently.

SOKO O NUKU TO IU KOTO
("Knocking Out the Bottom")

"Knocking out the bottom": When one fights an opponent and it appears on the surface that he has been defeated, if his fighting spirit has not yet been eradicated in his heart of hearts, he will not acknowledge defeat.

In that case, you must quickly change your mental attitude and break the opponent's fighting spirit. You must make him acknowledge defeat from the bottom of his heart. It is essential to make sure of that.

This "knocking out the bottom" is accomplished by means of the sword, the body, and also the spirit. It is difficult to be absolutely sure that you have succeeded. In cases where the opponent is completely defeated from the bottom of his heart, one does not need pay any more attention to him. However, when this is not the case, we must keep an eye on him. The opponent, if left with spirit, is difficult to defeat.

In both large-scale battles and single combat, one must diligently practice "knocking out the bottom."

ARATA NI NARU TO IU KOTO
("To Renew")

"To renew" refers to cases where you are engaged in combat with an opponent and you find yourself locked in a standoff. In this situation abandon your previous plan and with a spirit of starting afresh, get back into your timing and you will be able to discover the way to win.

On the occasions when you and your opponent find yourselves in a standoff, you must always simply change your plan and find another way to win. This is the meaning of "to renew."

Even in large-scale battles, it is important to understand this. If one has a sophisticated knowledge of *Heihō*, one can easily understand this matter. One must study diligently.

SOTŌ GOSHU TO IU KOTO
("Rat's Head, Ox's Neck")

The "rat's head, ox's neck" refers to the times when, while engaged in combat with an opponent, you attack each other, and paying attention to fine details, become deadlocked. In the path of *Heihō*, always think of *Sotō goshu* and suddenly switch from worrying about small details to concentrating on the big picture. Making judgments about changes in the situation is one of the aims of *Heihō*.

Bushi must always keep this rule in mind and apply it to everyday life. Forget this rule neither in large-scale battles nor in single combat. Study this diligently.

SHŌ SOTSU O SHIRU TO IU KOTO
("Commanders Know the Troops")

"Commanders know the troops" is a rule which in my doctrine of *Heihō* is always applied in time of battle.

Using your knowledge of military tactics, think of all the enemies as your own soldiers. Think that you know how to make the enemy move as you wish and try to move the enemy around freely. You are the general. The enemies are soldiers under your command. One must plan.

TSUKA O HANASU TO IU KOTO
("Letting Go the Hilt")

"Letting go the hilt" has several meanings. It means winning without a sword. Also, it has the meaning of being unable to win although carrying a long sword. I cannot fully record all the various shades of meaning. One must practice diligently.

IWAO NO MI TO IU KOTO
("The Body of a Massive Rock")

There is a thing called the "the body of a massive rock." By knowing the doctrine of *Heihō*, one, in no time at all, becomes like a massive rock. No one will be able to hit you. No attack whatsoever will disturb you. I teach this by word of mouth.

Epilogue of the "Fire Book"

These things written above are nothing but the things that I have constantly had in mind about the practice of the *Heihō* of the *Niten Ichi* school. This is the first time that I have recorded these teachings of winning in combat, so they may be out of sequence. I cannot break it down and discuss it in detail. Nevertheless, it is an absolutely necessary spiritual guide for the men who want to learn these teachings.

I have devoted myself to the study of the doctrines of *Heihō* since I was young, and in a general way turned my hand to swordsmanship. I have trained my body and disciplined my mental attitude in various ways. Furthermore, when I have examined the men of other schools, listened to their theorizing, and witnessed their applying themselves to perfecting their hand at fine techniques; although, at first glance, they look good, in fact not one of them has a sincere spirit.

Of course, these men too, although they are learning these kinds of things and believe they are improving the training of their bodies and perfecting the discipline of their spirit, these things, nevertheless, are all injurious to the way of *Heihō* and these bad influences never disappear. These things will become the cause of the decay of the true way of *Heihō* in the world and the decline of the Way of the Sword.

The true way of swordsmanship is to fight opponents and win the fight. Can it be anything other than this? If one masters my *Heihō* and wholeheartedly puts it into practice, there can be no room for doubt about you winning.

The twelfth day of the fifth month of the second year of *Shōhō*
[1645]

To Terao Magonojō

Shinmen Musashi

KAZE NO MAKI

("The Wind Book")

Commentary to the "Wind Book"
KAZE NO MAKI
("The Wind Book")

In this chapter, Musashi describes the various other schools existing at the time, and gives his views on their respective shortcomings. It is an important point that Musashi makes throughout: know your enemy.

You should read carefully, for here is presented the specific details of what makes Musashi's *Heihō* different from the mere swordsmanship techniques of the other schools. His discussion will give you an idea of what to look for in facing your own "enemies."

Do not allow the words to hold back your understanding. As words, they are inadequate. Musashi tells you this himself. Remember that this is to be your own experience. Let the words point you in a direction: you make the trip yourself. Open your mind to all possibilities. When you read "long sword," do not let your mind stop at "long sword." For Musashi, the tool was a shining steel blade. For a painter, it is the brush and the paint. For a lawyer, it is words, for a truck driver it is the truck and its schedule, and for the person in business it is the negotiating and arranging of events that make one successful. Let your creative mind understand, deeply and intuitively.

Introduction to the "Wind Book" KAZE NO MAKI ("The Wind Book")

Since it is important to be acquainted with the paths of the various schools, I shall write about them in this chapter which is called "The Wind Book." Unless one is familiar with the other schools, my *Niten Ichiryū* school cannot be fully mastered.

In investigating the other *Heihō*, I find some schools emphasizing the use of large long swords and placing particular importance merely on strength. On the other hand, there are some which specialize in the use of the *kodachi* ["short sword"]. And others have invented a large number of *tachi* ["long swords"], and teach in terms of how to take a stance, such as the *omote* ["external appearance"], and the *oku* ["internal appearance"], forms with the long sword.

This chapter will show why all of these are not the true path and will let the reader know of the good and the bad of these schools. The principles of my *Niten Ichiryū* school are special.

Each of the other schools look upon *Heihō* as one of the arts, and they wander off the true path in that they try to add polish in order to make theirs into merchandise which will sell since they look upon their work as a means of making a livelihood. There is in this world that which purports to be *Heihō*, which is limited to the practice of sword techniques; it attempts to find out how to win merely by practicing the swing of the long sword, by getting the body into good condition, and by polishing their techniques. In any case, such are not befitting of the true path.

Here, I shall write and explain wherein lie the inadequacies, one by one, of the other schools. One should examine this carefully and attain an appreciation and confidence as to the superiority of my *Niten Ichiryū* school.

TARYŪ NI ŌKI NARU TACHI O MOTSU KOTO ("The Holding of Large Long Swords in the Other Schools")

Among the other schools, there are some which are partial to large long swords. In my *Heihō*, this is considered a sign of weakness.

The reason for this lies in the fact that there is no understanding of the principle of "winning in whatever manner necessary" in the other schools, so that it is considered by those of the other schools that it would be desirable to win against an opponent from a distance by means of having an extra long sword, thus leading to the preference for extra long swords.

There is a popular saying, *issun te masari* [literally: "one-inch advantage"; figuratively: even an advantage of one inch is important], but such are for those who are ignorant of *Heihō*. Therefore, attempts to achieve victory over an opponent by means of the use of an extra long sword due to ignorance of the principles of *Heihō*, are manifestations of a weakness of spirit and are seen as the *Heihō* of the weak. If the distance between oneself and the opponent is short, so short that one is entangled with the opponent, the longer the long sword the more difficult it is to strike, and since it is also impossible in such situations to freely swing the long sword and the long sword becomes a burden, one is at a disadvantage then, in comparison to he who uses a short sword.

Although they who prefer extra long swords probably have their rationalizations for their preference, such rationalizations are subjective. Seen from the true path, these rationalizations have no grounds. Does one who holds a relatively short long sword always lose against one who holds an extra long sword? Also, when the situation is such that one is limited in space from above, below, and on the sides, or even those situations where one cannot use the short sword, the preference for an extra long sword is an indication of a deviation from the path, and is a wrong set of mind. There are some people who do not have much strength and the extra long swords are not meant for such people.

There is an old saying that "the greater embraces the less," and it is not that I dislike extra long swords for the sake of disliking them. It is just that I dislike the spirit which leads to the preference only for extra long swords.

Using an extra long sword may be compared to having a large army, and using a relatively short sword may be compared to having a small force. Is it impossible for a small force to win against a large army? There are many examples where a small force has been victorious over a large army. In my *Niten Ichiryū* there is a dislike for such a narrow-minded spirit. This must be carefully studied.

TARYŪ NI OITE TSUYOMI NO TACHI TO IU KOTO ("The Relatively Strong Long Swords in the Other Schools")

There ought not be any such thing as a strong long sword or a weak long sword. The long sword which is swung with a violent spirit is a coarse one. It is difficult to win only with brute strength.

Also, when one intends to cut another, if one strives only for the strength of the long sword and thereby tries to cut with unreasonable strength, it can happen that the reverse of one's intentions occurs and one cannot cut. In test cutting also, it is not a good idea to consider cutting with strength. In a cutting match with an opponent there is no one who considers trying to cut weakly or with strength. When you are trying to cut and kill someone, you are not thinking about cutting with strength, nor, of course, are you thinking about cutting weakly; rather, one is totally involved with getting the opponent to die.

And there is also the situation where one is using a relatively strong long sword and one strikes with much strength against an opponent; you lose your balance with the rebound effect of such a strong movement with a strong sword, which inevitably places you in a disadvantageous position. If you strike upon the long sword of the opponent with too much strength, the movements of your long sword are slowed down. From these reasons it can be seen that emphasis on the relative strength of a long sword is meaningless. Even in the case of *Heihō* as it pertains to large numbers, if one desires to have a strong army and to win with strength, the opposition will also endeavor to have a strong army and to win with strength. It is not possible to win in a fight unless one uses the correct principles.

The path of my *Niten Ichiryū* school does not in the least way concern itself with that which is impossible, but rather, it considers the means by which one can attain victory to be of the utmost importance. This ought to be thoroughly elaborated on.

TARYŪ NI MIJIKAKI TACHI O MOCHI IRU KOTO ("The Use of Relatively Short Long Swords by the Other Schools")

To win using only a relatively short long sword is not the true way. From ancient times a differentiation has been made between *tachi* ["long swords"] and *katana* ["swords"], which has been based upon the differences in length.

It is popularly thought that a strong man can wield a large long sword as if it were light, so there seems to be no reason why such a man would opt for a short sword if he had a choice. Spears and relatively long swords are used for their length. The mentality which would provoke one to cut, leap in, grab, and so forth, with a relatively short sword, in the space between the swings of the long sword of the other side is a distorted one and not good.

Trying to take advantage of an opponent's unguarded moment always places one in the defensive so that one is in a situation in which one can be led on by the opponent; therefore, I dislike such situations. Furthermore, attempting to get in among the opponents, with a relatively short long sword does not apply when one is dealing with a large number of opponents.

Those who use a relatively short long sword get tangled up since their long swords are always in the position of parrying; they are in a defensive position when they attempt to cut away against a large number of opponents or try to freely leap about. Such cannot be called a sure way. All other things being equal, it is important to achieve a sure victory by guarding oneself securely, by chasing around the opponents, by leaping and scattering them about, and by throwing them into a state of confusion.

The same principles apply when one is facing conflict against a large number of opponents. All other things being equal, it is the highest form of *Heihō* to aim with a large army for an opportunity when the opponents least expect it, and to promptly attack and crush the opponents.

If one, in his daily studies of *Heihō* practices only how to parry, return, dodge, and duck, such practices become a part of one's ways and it becomes likely that one will become prey to being led on by others. In the path of *Heihō*, because it is a straightforward and correct one, it is of the utmost importance that one have the spirit of chasing after and convincing others of the correct principles. This must be well researched.

TARYŪ NI TACHI KAZU ŌKI KOTO
("The Large Number of Long Swords in the Other Schools")

The reason why people are taught about the usage of a large number of long swords in the other schools is probably to make *Heihō* a salable commodity, since novices can be favorably impressed by knowing about a large number of long swords. This is something to be despised in *Heihō*.

The reason for this dislike of emphasis on the ways of using a large number of swords is that this is a symptom of wavering, although I do believe that there are many ways with which to cut someone. In reality, there is no substitute for cutting a person. For those who know *Heihō*, as well as for those who do not know *Heihō*, and for women and children as well, there are not that many ways of hacking down someone. In terms of the different ways, there is really no difference whether one thrusts or plows down the opponent. Since it is the way of *Heihō* to cut down the opponent, there are not that many ways by which to accomplish this.

Nevertheless, according to the locale and the circumstances, for example, there are the five methods which can be utilized in situations such as when the areas overhead

89

and to the side are blocked so that it is not possible to block with a long sword.

In addition, to cut down the opponent by twisting the arm or wrenching the body or leaping are not the true way. Twisting or wrenching do not serve any use when one is trying to cut someone down.

In my *Heihō* it is of primary importance to win by trampling down on the opponent, by confusing him, by twisting his spirit so that he loses his composure, all with a straightforward attitude and spirit. This should be well appreciated.

TARYŪ NI TACHI NO KAMAE O MOCHI IRU KOTO ("Making Use of the Long Sword Stances of the Other Schools")

It is a mistake to think only of the long sword stances. Positions, as they are popularly understood, are the positions which are taken when there is no opponent.

This is because in the path of *Heihō* one cannot be so rigid as to insist that "this is the way it has been since ancient times," or "this is the modern way to do it." The path to victory lies in manipulating the circumstances so that they are to the disadvantage of the opponent.

A stance is not that which changes according to the circumstances, but rather the precaution with which one assumes a firm and steady position. When one speaks of stance in terms of a castle, one is referring to the condition in which it is impregnable if and when it is attacked. In the path to victory in *Heihō* taking the initiative at all costs is the most important thing. To be in a stance is to be in a state of waiting for the initial attack by the other side. There is the need to carefully consider this.

I dislike the defensive spirit which is inherent in a stance. One must irritate the stance of the other side, by designing strategies which the opponent would never foresee, by confusing, upsetting, or frightening the opponent, or by agitating him so as to break his rhythm. Thus it is said in the path of my *Niten Ichiryū* school that "there is and there is not a stance."

In *Heihō* as it pertains to large numbers, it is of the utmost importance to initiate the conflict with a strategy which has been designed to make the best of the advantages presented, in terms of the number of opponents, the conditions of the site of conflict, and the strength of the men on one's own side. It is twice as difficult when the initative is taken by the opponent as compared to when you take the initiative. To take a good stance with the long sword, and to be well prepared to "fend off," are the same as if one were a fence which is immovable against long objects such as spears and extra long swords. To look at this from another perspective, it is important to have fervor such that one would be willing to pull out a fence railing and use that in place of a spear or an extra long sword. This is such that it must be well and carefully studied.

TARYŪ NI METSUKE TO IU KOTO
("The Points of Concentration of the Other Schools")

Depending on the school, one fixes one's point of concentration on different things; according to some on the opponent's long sword, and according to some others on the hands. There are also some which advocate that one fix one's point of concentration on the face or the feet.

But if one attempts to fix one's point of concentration on one thing, the result will be a wavering of the spirit and a hampering of *Heihō*. For example, a player of Japanese football is capable of various and difficult kicks even when he has not fixed his point of concentration on the ball. He does not have to actually see the ball itself since he is accustomed to balls. A trained juggler can stand a sliding

91

panel door on his nose or juggle several swords. He can perform such acts without fixing his point of concentration on the objects which he is balancing or juggling because he is accustomed to it and is capable of seeing them without concentrating on them.

In the path of *Heihō*, one can see well even the distance and the swiftness of the opponent's sword spontaneously if one is accustomed to combat with individual opponents and knows the ups and downs of the opponent's spirit, and has mastered the path. In *Heihō*, one may generally consider the spirit of the other person to be the subject of the point of concentration.

Also, regarding the point of concentration in terms of *Heihō* as it pertains to large numbers, it is necessary to be able to see the number of the opponents. As for "observing and seeing," it is important to strengthen the eyes with which one observes in order to be able to read into the opponents' mind, to see the area, to grasp the overall situation, to judge which side is doing better, to consider the strengths and weaknesses of both sides at any given time, and finally, to assure victory.

In *Heihō*, as it pertains to both large and small numbers, it is not necessary to observe minor matters. As I have stated before, if one fixes one's point of concentration on minor matters, one loses sight of the major concerns, and this in turn creates vacillation of the mind and causes one to lose opportunities for a sure victory. Understand this process and train well in this perspective.

TARYŪ NI ASHITSUKAI ARU KOTO
("Footwork in the Other Schools")

There are many types of footwork, such as the light step, the flying step, the jumping step, the stomping step, the side step, and so on. All are unacceptable in my *Heihō*.

I dislike the light step because it always makes your step light in combat, just when you need a firm step the most.

I do not like the flying step because it is easy to get into the habit of it. The flying step is not good because there is no need to take the flying step repeatedly.

The jumping step is based upon the intention to leap off, so it is executed very quickly. The stomping step is also called the waiting step. I particularly dislike this type of footwork. There are also other types of fast footwork, including the side step.

There are places such as muddy bogs, streams, stony fields, and narrow lanes where it is impossible to jump around and use fast footwork if one wishes to be able to cut the opponent. There are no differences in the footwork in my *Heihō*. No distinction is made from the steps which are used when walking down a common road. I follow the rhythm of the opponent, assuming the body posture which is appropriate when he hurries or when he slows down, keeping my footwork under control, and without allowing it to become too fast or too slow.

Footwork is also important in *Heihō* as it pertains to large numbers. The reason is that if one hastens to attack without knowledge of the opponents' strategy, one loses one's rhythm and cannot win. Also, footwork which is too slow prevents one from taking advantage of the opponents' confusion and disequilibrium, so that one loses one's opportunity to win, and thus, one cannot expect to achieve a quick victory. It is essential to judge when the opponents are confused and out of control, and not to give them any opportunity to relax. Train well in this.

TA NO HEIHŌ NI HAYAKI O MOCHI IRU KOTO
("The Concept of Speed of the Others")

In *Heihō*, speed is not the true way. Speed is the fastness or the slowness which occurs when the rhythm is out of synchronization.

The movements of a master of a path do not appear to be unduly fast. For example, there are runner couriers who run as much as ten to twelve miles in a day. They do not run at such great speeds from morning til night.

An unskilled runner courier may run all day, but he will not cover a great distance. In the path of song and dance, when an inexperienced singer is teamed with an experienced one, the inexperienced one feels that he is lagging behind. When an inexperienced drummer tries to beat out the tune of a slow song such as *"Rōmatsu"* on a drum, he finds himself feeling that he is too slow. There are fast songs such as *"Takasago,"* but these too should not be played too quickly. As the saying goes, "Haste makes waste," and the timing is upset when one proceeds too quickly. Of course, too slow is also bad.

What a master does seems to be done with ease and without any loss of timing. Anything which is performed by someone who has experience does not look busy. Learn about the principles of the path from these examples.

In the path of *Heihō* in particular, speed is bad. The reason is that some places, such as swamps, do not allow swift body and foot movements. Since a long sword cannot be wielded like a fan or a small sword, you will find that you cannot cut at all if you try to cut swiftly under such circumstances.

It is also not good to make haste in *Heihō* as it pertains to large numbers. Making as if holding a pillow down will help you so that you will not be in the least too slow.

Also, if the opponent is hastening, it is important to resist this tendency and to remain calm so as to avoid being manipulated by the other side..One should train in how to make use of this spirit.

TARYŪ NI OKU, OMOTE TO IU KOTO
("The Concept of Inner and Outer in the Other Schools")

In matters which pertain to *Heihō*, all things are outer and nothing is inner. In some of the arts, people sometimes speak of that which is top secret, which is only for the insiders, and that which is for novices, but in actual combat situations no one fights with an outer technique or cuts with an inner technique.

When I teach my *Heihō* to a beginner, I get him to practice and learn those techniques which he can learn quickly, and as for those points which are difficult, I ascertain his readiness and then teach the deeper principles gradually. Nevertheless, since most of my teaching is concerned with knowledge which is acquired from actual experience, the differentiation between beginning techniques and advanced techniques is not stressed.

Thus, if one is to go deeper and deeper into a mountain, one will eventually find oneself outside at the entrance. In all paths there are some things which can be learned only as one delves deeper and deeper. And there are things which can be understood by a novice. That is why I do not present certain principles of combat as being secret and others as being open. Thus, in my way, I do not like the idea of oaths and penalties. The path for teaching my *Heihō* is to judge the intelligence of the students of the path, and based on that, to teach the true path, to get them to rid themselves of the shortcomings of the various other schools of *Heihō*, to help them to perceive on their own the true path of the Way of one of the warrior class, and to make their hearts above suspicion. One should train well in this.

Epilogue to the "Wind Book"

So far, I have written in the Wind chapter descriptions of the *Heihō* of the other schools in the above nine sections, and although one should present explicitly the teachings of each school from the basic to the advanced principles, I have purposely avoided mentioning names of schools and their techniques.

It is because each school has its own opinions, each path its own claims, and also because there are differences of opinion from one person to another, even within the same school; I have not mentioned the schools for the sake of the future.

I have placed each of the other schools into nine categories, and if the popular ways are examined with the correct principles, it can be seen that those which lean toward the long, those which use the short as their principle, those which lean toward the strong or the weak, and those which are attuned to the overall picture or to the fine details are all distorted and prejudiced, so that even though I have not actually given the elementary and the profound techniques of these other schools, it should be obvious to all. In my school there is no concept of that which is the elementary and that which is the profound regarding the long sword. There are no secret stances. The most important thing in this *Heihō* is to keep in mind the value of its ideas.

The twelfth day of the fifth month of the second year of *Shōhō*
[1645]

To Terao Magonojō

Shinmen Musashi

KŪ NO MAKI

("The Book of Emptiness")

Commentary to the "Book of Emptiness"
KŪ NO MAKI
("The Book of Emptiness")

Thirty spokes will converge
In the hub of a wheel;
But the use of the cart
Will depend on the part
Of the hub that is void.

With a wall all around
A clay bowl is moulded;
But the use of the bowl
Will depend on the part
Of the bowl that is void.

Cut out windows and doors
In the house as you build;
But the use of the house
Will depend on the space
In the walls that is void.

So advantage is had
From whatever is there;
But usefulness arises
From whatever is not.

The Tao Te Ching

What follows is a brief philosophical discussion to
prepare you for the final chapter of *The Book of Five
Rings,* "The Book of Emptiness."

Philosophers have forever been inquiring into the nature of reality. Basically, the arguments have been over what is real, that is, what can be said to exist. For instance, we might all agree that this book exists; you can see it and touch it. But what about "beauty"? Does "beauty" exist in the same way as this book? No, you can see or touch a beautiful *thing*, but not "beauty" itself. Would it be correct to say then that "beauty" does not exist? No, because when we talk about it, we all know or have some idea about it, so it exists somehow. So if it does exist, where is it? Some say it only exists in your mind, that "beauty" is only a human conception, a human description of "real" things, used in the same way as when we describe something as "red."

Maybe it would be safer to say that "beauty" exists in a different way than this book does. Well then, let us consider chiliagons. Picture a chiliagon in your mind. What is a chiliagon, you say? You mean you don't know what it is, that you can not picture it? Does that mean that it does not exist?

A chiliagon is a thousand-sided figure. Now picture it. Still having a problem? But now you know what it is! Does it exist yet? You might be more sure if we were discussing a three- or even a ten-sided figure, because that is easier conceptually. But this chiliagon is impossible to envision. That does not mean that it does not exist, though. With enough space you could build one. Now relate this to the idea or concept of "beauty." Is "beauty" more real than the chiliagon? In your mind's eye you can "point" to something "beautiful," but you cannot "point" to a chiliagon.

How about unicorns? Can you picture a unicorn? You probably can, yet there is no such thing as a unicorn, is there? So which is more real, the chiliagon that you can not conceive of but is possible to build, or the unicorn that you can picture but that you know you will never see?

Consider also the following: If you can see red and green, you know those colors "exist." For a severely color-blind person, who might not be able to see red or green colors, "red" and "green" do not exist. Now we are discussing the subjective experience as a criterion for existence. Experiencing it with your senses tells you if it is real or not.

The *Zen Buddhist* view of reality is neither subjective nor objective. In it there is no distinction between the "I" and the "that." Neither consciousness (your own subjective experience) nor external reality (objective, having nothing to do with the experiencer, for example, "beautiful-ness") is considered "ultimate." The *Zen* view, as always, is a practical one. These considerations are irrelevant. Everything is real only in relation to other things. Everything is devoid of reality, that is, nothing has its own *independent* existence. Consciousness and external reality are both real only in relation to each other. Or to put it another way, all that exists, exists relative to other things.

The traditional *Zen Buddhist* view is that the world is in a constant state of change. Nothing in the world ever "is"; it is always "becoming." The nature of reality is a process, a continuous changing flow. Nothing is absolute; it is always changing. This process, this flow, is *kū,* ("emptiness").

In the *"Shingyō Sutra"* chanted by the *Zen* monk each morning appears the following:

Form is here emptiness, emptiness is form; form is no other than emptiness, emptiness is no other than form; that which is form is emptiness, that which is emptiness is form.

It is the relationship that "is," not the things that relate. This relationship is *kū.*

Musashi says, "That which can not be known is *kū* ("emptiness")." This emptiness is not a mathematical void; it does not mean nonexistence, that something does not exist. Neither is it something unknowable, in the sense of "inexperienceable." Emptiness is meant to describe something which is "devoid" of an *independent* reality. This "something" is in the nature of relationship, that is, a process, a flow, in which all things are in a continual state of change.

It is this process that is *kū*. It is the realization of this process that the *Zen* practitioner is trying to attain. Experiencing this flow, this process, directly, personally and immediately is what is meant by "being one with the moment."

Understanding that the nature of reality is process implies that one has gone beyond the point of making distinctions, creating dualities. When you make these distinctions, your mind "stops," and you are no longer aware of, or more critically, no longer in tune with this flow.

To posit "beauty" or "book" or "unicorn" or "chiliagon" is to have your mind stop. To think of death when you are faced with your enemy is to have your mind stop. This is why the swordsman must remain detached from "worldly" thoughts. *Munen musō* takes on new meaning. If you can rid yourself of the "stopping mind," you will achieve *Satori,* and experience the moment as if it were your own.

You have your own flow, your own process. When Musashi suggests that you "make *kū* your path, and your path as *kū*," he is suggesting that there is a higher order of experience than the one you are on now. The emptiness is really a fullness, the realm of all possibilities. What Musashi and the *Zen* practitioners are saying is that via *Heihō* or whatever path you follow, one can attain an understanding of his own place in the process. If your mind is open, you are free to be with the flow, to be in rhythm with the timing of change.

Musashi has attained that. With a proper spirit, an honest heart, a dedication to what is just, a benevolence that is boundless, and an iron will to succeed, he says that you can do it too.

KŪ NO MAKI
("Emptiness")

The path of *Heihō* of *Niten Ichiryū* is given in this *kū* chapter.

The meaning of *kū* is emptiness; that which cannot be known is *kū*. Of course, *kū* is emptiness. By knowing form, one knows emptiness. This, in short, is *kū*.

People in general, when they cannot understand something, call it *kū*, but this is not the true *kū*. This is illusion.

When practicing the path of *Heihō* as a *bushi*, if one does not understand the way of the *bushi*, this is not *kū*. When one has been led astray or filled with false conceptions, and cannot resolve these problems, calling this *kū* is not *kū* in its true meaning.

For a *bushi*, knowing the path of *Heihō* with certainty, acquiring skill in the other martial arts, understanding clearly the road to be followed by the *bushi*, having no illusions in your heart, honing your wisdom and willpower, sharpening your intuitive sense and your powers of observation day and night; when the clouds of illusion have cleared away, this is to be understood as the true *kū*.

While one is ignorant of the true path, he is convinced that his path alone is correct, whether he believes in the path of Buddha or any other teachings of the world. But when seen from the standpoint of the true path and in the light of the situation in the real world, these views divert from the true path because of individual prejudices and distorted points of view.

Taking this into proper consideration, with a straightforward spirit as your foundation, and an honest heart as your path, practice *Heihō* broadly. It is important to judge the general situation clearly and correctly. Make *kū* your path, and your path as *kū*.

In *kū* there is good, and there is no evil. When there is wisdom, reason, and the Way, there is *kū*.

The twelfth day of the fifth month of the second year of *Shōhō*
[1645]

To Terao Magonojō

Shinmen Musashi

ABOUT THE TRANSLATORS

BRADFORD J. BROWN is Executive Director of the Nihon Services Corporation. Trained and educated as a lawyer, he has been providing counseling and assistance to the Japanese community in New York for many years. A student of Zen for more than half his life, he is associated with the Shorinji Kempo training hall in New York City.

YUKO KASHIWAGI, Senior Staff Interpreter, has been a professional interpreter and translator of Japanese and English for over 12 years. Her family has been translating and interpreting English and Japanese for two generations in Japan. Educated in both Japan and the United States, she is a graduate of International Christian University in Tokyo.

WILLIAM H. BARRETT, Senior Staff Interpreter, has been interpreting and translating Japanese for over a decade. He has attended universities in both Japan and the U.S. and has earned three graduate degrees at Columbia University in Japanese history and language. He has been a language officer with the U.S. State Department for over ten years.

EISUKE SASAGAWA, Senior Staff Interpreter, has been a professional interpreter and translator of the Japanese language for ten years, working in Europe and the Middle East as well as the United States. He is a U.S. State Department certified Escort Interpreter. Along with Japanese and English, he speaks German, Hebrew and Yiddish.

Nihon Services Corporation is an interpreting, translating and business counseling service based in New York City. It is dedicated to breaking down cultural and communication barriers between Japan and the United States.

MONEY TALKS!
How to get it and How to keep it!

YOU CAN TAKE ADVANTAGE OF THE FINANCIAL OPPORTUNITIES AVAILABLE TODAY USING THESE FINE BOOKS FROM BANTAM AS GUIDES

We Deliver!
And So Do Those Bestsellers.

Special Offer
Buy a Bantam Book
for only 50¢.

Now you can have Bantam's catalog filled with hundreds of titles plus take advantage of our unique and exciting bonus book offer. A special offer which gives you the opportunity to purchase a Bantam book for only 50¢. Here's how!

By ordering any five books at the regular price per order, you can also choose any other single book listed (up to a $5.95 value) for just 50¢. Some restrictions do apply, but for further details why not send for Bantam's catalog of titles today!

Just send us your name and address and we will send you a catalog!